Islands

Islands

by John Fowles

Photographs by Fay Godwin

LITTLE, BROWN AND COMPANY
BOSTON TORONTO

The photographs are for Rex and Zelide Cowan,
with love, and gratitude for all their help and hospitality.

Library of Congress Catalog Card No. 79-59165

First American edition

Printed in Great Britain

Preface

The course of my affections, in terms of love for the landscapes of Britain, has always been very firmly set south-west, and above all for those two counties whose lifeblood has always been the sea, and who point, like a massive arm bicepped by Devon, forefingered by the Land's End peninsula, to some secret centre further south-west still – an image that may seem farfetched today, but was certainly not so to the original Celtic and pre-Celtic inhabitants of the region. I still feel a lift of something more than machinery when the helicopter rises from the ground at Penzance and heads westward, even though I know I shall be landed on my least favourite of the Scilly Islands, and shall once again disapprove its bungalows, its echoes of the mainland, its overstayed tourists trying to be more local than the locals – all its insularity in the pejorative sense. I am being a good deal less than kind to St Mary's, which does have fine walks and pretty countrysides, and whose faults are minute compared to those of the other contemporary hell-holes of the extreme West.

In fact, all the islands are under the protection of the Duchy of Cornwall and even more importantly, under that of a strong native disinclination (which makes them rare indeed these days) to give any ground to the dominant syllogism: profit is life, tourists bring profit, therefore let us sacrifice everything to tourism. None the less, the true uniqueness of the Scillies is best seen in the eight non-tourist months of the year. It is no good going to Looe, Polperro and St Ives then; they have been exploited and vulgarized beyond any hope of redemption by season – in winter they are merely dormant cancers. It is very different in the islands. They can be wet, they can be wild, they can be misty and all the things (except cold) that travel agents fear: but they are most beautiful then, and certainly most capable of communicating the things I shall be writing of. It is when to enjoy them most, to *feel* them most. Then they remain a kind of omphalos, a Delphi of the South-West. One returns to the roots of something beyond one's personal descent.

Even when I was writing the text of a very different book on the Scillies, *Shipwreck*, I knew I wanted to analyse this attraction to something rather more than just place, though place was inherent in it. But I could not at the time think of a photographer who saw the bare bones of sea and land in the way that what I had in mind required. Then by chance Fay Godwin, in pursuit of the bread-and-butter side of her profession, came to take some shots of me. She mentioned that she had done the photographs for a guide to the Ridgeway, and sent it when it was published. At the first page I opened I knew I had found a partner. British photography has not had a more poetic interpreter of ancient landscape, of its lights and moods and forms, for many years; and I doubt a more painstaking one. One day in her studio I said of a print (it can be seen on the cover of *The Drovers' Roads of Wales*) that she had been lucky to catch such a perfect sky. 'I didn't catch it,' came the deservedly tart reply. 'I sat down and waited three days for it.'

Her art speaks the purest and most direct appeal of the archipelago much more eloquently than words could ever achieve, or certainly my words – which is as it should be, since we began with the firmly shared view that we should interpret the basic theme individually and separately, and agreed that in this case photography should finish before writing began; which made my side of it much easier.

There was another kind of book I never intended to write – a guide to the Scillies. Guides and histories of the islands abound. The best of the first remains, in my opinion, Geoffrey Grigson's *The Scilly Isles* of 1948; of the second, R. L. Bowley's *The Fortunate Islands*. In any case, and even if these and countless more specialized studies did not exist, I am not in any way an expert on the Scillies. I have never lived there, I can count my visits on one hand. This is much more about the Scillies of a novelist's mind; and beyond them, about the mysteries, symbolic and real, of all similarly situated small islands; about their silences, their otherness, their magi and their mazes, their eternal waiting for a foot to land.

I must thank Penguin Books for permission to quote from E. V. Rieu's version of Homer's *Odyssey*; my fellow-curator at the Philpot Museum of Lyme Regis, Ann Jellicoe, and her husband, Roger Mayne, for archival help and for pointing out the Kilvert references; and especially thank Mr Llewellyn Vaughan-Lee and Dr Malcolm Evans for allowing me to read in typescript their paper 'The Vernacular Labyrinth: Mazes and Amazement in Shakespeare and Peele', from which I have shamelessly borrowed both the central thesis and some of the evidence for it. Their discussion is far more scholarly and detailed than mine, and it will be published in the 1979 yearbook of the German Shakespeare Society.

J.F.

Islands

The wise visitor to the Scillies does not drive straight to Penzance and board a helicopter or a ship, but finds time, so long as the weather is clear and visibility good, to go out first to Land's End. And there they float, an eternal stone armada of over a hundred ships, aloofly anchored off England; mute, enticing, forever just out of reach. The effect is best later in the day, when they lie in the westering sun's path, more like optical illusions, mirages, than a certain reality. I say 'they', but the appearance at this range is of one island; which has a justice in it, since in remote antiquity all the larger islands except St Agnes very probably were conjoined.

At Land's End you already stand on territory haunted by much earlier mankind. Their menhirs and quoits and stone lines brood on the moors and in the granite-walled fields; and even today the Scillies can in certain lights lose the name we now call them by and re-become the Hesperidean Islands of the Blest; Avalon, Lyonesse, Glasinnis, the Land of the Shades; regain all the labels that countless centuries of Celtic folklore and myth have attached to them. Adam and Eve braved the sea, probably as long as four thousand years ago. Their burial places are scattered all over the present islands, and so densely in places that one suspects the Scillies must have been the ultimate Forest Lawn of megalithic Britain, though interment there would not have been an ambition of only the dying. The spirits of the dead could not cross water, and the living may well have cherished that thirty-mile *cordon sanitaire* between themselves and their ancestors. Whatever the reason, the islands hold an astounding concentration of nearly one-fifth of all such tombs in England and Wales – far more than Cornwall, which is already rich in them.

Some of the great boulders, naturally carved by Atlantic wind and rain, split and isolated by the Ice Age, that the earliest settlers found there would have profoundly impressed, and baffled, them. They are so splendidly wrought and monumental – especially on Gugh and the south side of St Agnes – that it is as if some earlier incarnation of Henry Moore has played a huge joke (in one case a huge phallic joke) on posterity. The pluperfect one lies on the furze-moor just above Porth Askin, exquisitely posed and pedestalled in a rainwater pool. It would grace the forecourt of any

twentieth-century skyscraper: and much higher praise, not disgrace the most fastidious Zen garden. Perhaps it was these magnificent stones that seeded the legend of the lost land of Lyonesse and the associated myth of Atlantis; of a simpler, nobler, vanished world and culture.

There is a more likely origin of the legend. The ancient Celtic inhabitants of Cornwall and the islands almost certainly had contacts with, if not a nobler, at least a more advanced culture — whose ships would have appeared out of the south-west, even though their homeland lay in quite another direction. The Phoenicians were the great trading, exploring and sea-going race of antiquity. According to Strabo they had discovered the Atlantic before 1000 B.C. Their colony at Cadiz dates from about that time. By an irony they are both the most commercial and the most mysterious of ancient civilizations — mysterious because they left so few traces of their existence. Their barter-currency presumably lay most in perishable goods, which makes them the despair of archaeologists. What is quite definitely known is that they coveted tin, which they used not only as a metal but as a dye mordant (stannous chloride); and that they regarded its British source as one of their most precious trade secrets. The Phocian Greeks who colonized Marseilles about 600 B.C. discovered it at some later date; and Herodotus knew the tin came from islands called the Cassiterides (Greek *kassyo*, to stitch together, and *kassiteros*, tin), but otherwise only that they lay somewhere very remote in Northern Europe. The first coherent account is by Diodorus Siculus, writing in the first century B.C.

The inhabitants of that part of Britain which is called Balerium (Land's End) are very fond of strangers, and from their intercourse with foreign merchants, are civilised in their manner of life. They prepare the tin, working very carefully the earth in which it is produced. The ground is rocky, but it contains earthy veins, the produce of which is ground down, smelted and purified. They beat the metal into masses shaped like astralgi (dice) and carry it to a certain island lying off Britain called Ictis (St Michael's Mount).

Strabo's report, from about the beginning of our era, runs as follows:

The Cassiterides, opposite to the West Parts of Britain, situate as it were in the same climate with Britain, are ten in number and lie near each other, in the ocean toward the North from the haven of Artabri. One of them is desert, but the others are inhabited by men in black cloaks, clad in tunics reaching to the feet, girt about the breast, and walking with staves, thus resembling the furies we see in tragic representations. They subsist by their cattle, leading for the most part a wandering life. Of the metals, they have tin and lead, which, with skins, they barter with the merchants for earthenware, salt, and brazen vessels.

Artabri is near Cape Finisterre, the north-western land's end of the Iberian peninsula. Both Pliny and Solinus, writing a little later, confirm the identifications of the Cassiterides with the Scillies. Tin-smelting pits that can be dated to 300 B.C. have been found near St Just. It seems probable that the tin on the Scillies themselves was always more exposed and easily exploited, even largely exhausted by Roman times. But the island metal was still mined in the sixteenth century, and found in workable quantities as late as the eighteenth; and no doubt the islands remained a depot for the mainland 'exporters' long after local supplies ran out.

In short, though positive proof is lacking, there does seem strong circumstantial evidence to suggest that mysterious strangers were descending on the extreme south-west of Britain, and regularly, from at least Homer's time and possibly even before it. I believe myself that this is where the northern version of the Atlantis corpus of legends springs from. A comparison with the kind of myths that the conquistadors in America, or the first explorers in Polynesia, gave birth to is illuminating. Man has never liked rational explanations of why strangers are more intelligent and technologically advanced.

I should have made a very poor hand on Ulysses' boat, since I have never in my life had to go by an island without wishing that I could have landed on it, and even in less than traditionally romantic circumstances. I had the longing only very recently, on a tour round Manhattan; would have had our launch stop at all those forsaken islets with their dilapidated warehouses and weed-jungles. In some way they put to shame the far more famous island they surrounded, and remained of their kind, where it has become a termite-heap. True islands always play the sirens' (and bookmakers') trick: they lure by challenging, by daring. Somewhere on them one will become Crusoe again, one will discover something: the iron-bound chest, the jackpot, the outside chance. The Greek island I lived on in the early 1950s was just such a place. Like Crusoe, I never knew who I really was, what I lacked (what the psycho-analytical theorists of artistic making call the 'creative gap'), until I had wandered in its solitudes and emptinesses. Eventually it let me feel it was mine: which is the other great siren charm of islands – that they will not belong to any legal owner, but offer to become a part of all who tread and love them. One's property by deed they may never be; but man long ago discovered, had to discover, that that is not the only way to possess territory.

It is this aspect of islands that particularly interests me: how deeply they can haunt and form the personal as well as the public imagination. This power comes primarily, I believe, from a vague yet immediate sense of identity. In terms of consciousness, and self-consciousness, every individual human *is* an island, in spite of Donne's famous preaching to the contrary. It is the boundedness of the smaller island, encompassable in a glance, walkable in one day, that relates it to the human body closer than any other geographical conformation of land. It is also the contrast between what can be seen at once and what remains, beyond the shore that faces us, hidden. Even to ourselves we are the same, half superficial and obvious, and half concealed, labyrinthine, fascinating to explore. Then there is the enisling sea, our evolutionary amniotic fluid, the element in which we too were once enwombed, from which our own antediluvian line rose into the light and air. There is the marked individuality of islands, which we should like to think corresponds with our own; their obstinate separatedness of character, even when they lie in archipelagos.

It is not only the geologists and ecologists who feel that in the Scillies: the islanders do themselves. In the old days there were different nicknames for the men of each island. Those of St Mary's were Bulldogs; of Tresco, Caterpillars (perhaps moonlit files of smugglers); of Bryher, Thorns (all thorn trees on the Scillies are blown askew, and Bryher people were supposed to look 'lopsided'). St Martin's men were Ginnicks, a word whose meaning R. L. Bowley says is lost, though I see Joseph Wright has it down (admittedly from a county on the opposite seaboard of England) as a synonym for 'neat'. On St Agnes you were a Turk, because you looked Spanish – the imputation being against the women of that island. St Agnes is nearest the dreaded Western Rocks and has given first shelter to more shipwrecked sailors than all the other islands together. It must therefore have seen Spanish Armada men, who everyone in the South-West knows may have been failures at sea, but were terrors in bed. I have a 'Spanish' great-aunt and uncle myself, so I believe every word of it. 'Turk' simply means not English; that is, outlandish.

Despite the much greater intercommunication and intermarriage of modern times, this separatism, or patriotism, has not quite disappeared. The respectively most 'foreign' and most 'native' of the present five inhabited islands, Tresco and Bryher (though St Martin's might claim the latter distinction), lie not much more than a long stone's throw apart. I sat once with a young Bryher wife. She came originally from Bristol, but she talked about the mainland not seeming 'real' any more. Then she looked out

of the window and across the narrow channel. 'Even Tresco,' she said; and I remembered Armorel, of whom more in a minute. Armorel felt rather the same about Bryher, which in another direction lies only a slightly longer stone's throw from her own island of Samson.

Island communities are the original alternative societies. That is why so many mainlanders envy them. Of their nature they break down the multiple alienations of industrial and suburban man. Some vision of Utopian belonging, of social blessedness, of an independence based on cooperation, haunts them all. Tresco is leased and managed by the Smith family, who have generally brought in outsiders to work there. I asked another native of Bryher what he thought of Tresco. He spat over the lee gunwale of his boat, which may seem ungrateful, in view of all the Smiths become Dorrien-Smiths have done, and are still doing, for the economy and conservation of the Scillies as a whole; but the spitting was, I knew, not against man, but against principle. He was prepared to make most of his summer living ferrying holiday-makers to the place; he allowed the charms of its modern hotel, its jolly pub, its famous subtropical gardens; but he would leave the Scillies sooner than live on Tresco himself. And he used finally a phrase that was almost one of pity, as if speaking of a fat girl trying to be a ballerina. 'It's not island,' he said.

Of course all islanders have to be handy with boats, but genuine nesomanes are not sailors. Centuries of professional mariners have wished the entire Scillies sunk a hundred fathoms deep, and most other small islands with them. Yachtsmen may enjoy archipelago-cruising, but their true marriage is to the moving island under their feet, whereas the attachment of the fanatic is to all that the passing craft will never know. The surrounding sea is an indispensable part of the setting for this obsession, but it is not of its essence. It is the isolater, not the isolation. The ships it carries on its horizon are like arrows that have missed the target, space vehicles heading for some other planet. They may photograph the surface, but will never know the interior.

However, it is not for nothing that since remotest antiquity the domain of the siren has been where sea and land meet; and it is even less for nothing that the siren is female, not male. Something deeper than aboriginal sexual chauvinism – the fact that ship-handling was always a man's affair – lies behind this. It is odd, if you think, that Ulysses should feminize both his bulwark and its age-old greatest enemy, the reef, the rock, the uncharted shore; almost as if being wrecked was the result of a quarrel between women, eternal destroying Scylla versus eternal launching Helen of Troy.

I think we must read here a paradox about possession, or possessibility.

Good sailors have always married their ships as well as their wives. A recent book showed that even those sailors in stone ships, offshore lighthousemen, can still enter into a very curious emotional relationship with their towers; not at all the sort of thing that the economic view of man as mere smoke in the wage-labour wind allows for. But of course to possess is always to want to possess more. No earlier man ever went voyaging for fun, or risked the sirens for the sheer hell of it; he went to find land, food, tin, gold, trade, *Lebensraum* . . . power of some kind. His voyage was done in what he already possessed, though never as securely as he wanted, and towards what he wanted to possess in addition, though never as certainly as he imagined. Aeons of empty marriages to pretty faces lie behind the siren; Adam may have delved in the literal earth, but he never scratched much beyond the surface of Eve's mind and nature. It was she, after all, who provoked the very first voyage, outward bound from Eden with her dimwit of a husband.

The historically very abrupt discovery of the charms of the seaside has always seemed to me one of the most bizarre happenings in the cultural history of Europe. Before 1750 it is almost as if everyone felt about coasts as many people feel today about airports. Of course one goes to them if one has to travel by air; of course one lives by them if one's livelihood depends on it or other circumstances oblige; but who in their right mind would ever go to an airport from choice, just for idle pleasure? The analogy may seem absurd, but that makes our apparent long blindness to the very real pleasures of the seaside even more mysterious. It is true that well up to 1700 most foreign tourists descending on European beaches came with cannon trained and cutlass in hand. The last place to have spent a happy holiday in the summer of 1690 was on the Dorset or Devon coast. Admiral Tourville's French fleet spent most of it cruising close inshore looking for towns to sack. Teignmouth was burnt to the ground, and other places bombarded. And it is also true that the very notion of 'holiday' was, in all but its original and literal sense, for most of the world a Late Victorian invention. Yet a mystery does remain: how can something so nice have been ignored for so long?

The change came, like most major human changes, from a conjunction of two factors. Man can follow reason against pleasure, and pleasure against reason, but when the two combine they are irresistible. In this case the medical profession and the first Romantics spoke as one. The doctors discovered the medicinal value of sea-bathing – and even sea-drinking, in the early days. The Romantics discovered nature and the picturesque. The sea was therefore passed good for both body and aesthetic soul; for me, in a word. The annual conferences of the Amalgamated Union of Sea-Sirens must have been gloomy affairs in the early eighteenth century; everywhere they were being declared redundant by the new lighthouses, the improved navigational methods. Then suddenly miracle: they had a brilliant idea. Instead of combing their tresses and facing out to sea, they would comb their tresses and face inland. Instead of corrupting sailors, they would pervert the landlubbers.

This abrupt acquisition of new victims can be very accurately dated in the case of my own town of Lyme Regis. There was a third factor involved: international politics. Again and again during the first decade of the eighteenth century Lyme was beseeching Her Majesty in Council, and the Duke of Marlborough, to send cannon and powder against the 'insults of enemy privateers'. Then there are thirty years of silence on the matter. In 1740 John Scrope was sent by the Privy Council to inspect the town and reported '. . . by the long peace their Fire Arms have been so much neglected that upon a late inspection of them there was not a Musquet in the Town that could be fired, so that the town is in a neglected and defenceless condition.' Six nine-pounders were sent, so Lyme was no longer defenceless; neglected it remained, but in that 'long peace' salvation had been brewing.

By 1750 the place was moribund, a warren of hovels, with all but two of its former opulent medieval and Tudor houses in ruins. Its harbour was in decay, an early victim of gigantism, far too 'tight' and shallow for the merchant ships of the time. Its one other ancient industry besides sea-trading, serge-weaving, was being throttled – as everywhere else in the West of England – by the better organized and more competitive North of the country. Nobody ever visited it; or easily could visit it even if they wanted to, since there was not a single carriage road. It was as near dead as one can imagine; and our coasts were sick with hundreds of towns in an exactly similar state.

But in 1770 an alert, farsighted and exceptionally generous man descended on the corpse. His name was Thomas Hollis, the benefactor of Harvard, and early socialist before the name – a radical, a *philosophe*. He told Lyme that its only hope was to make itself a little more presentable, and showed it the way to start. He bought the hovel property in the town not to profit from it, but simply to have it sledge-hammered to the ground. He cleared a little central square (now lost again, such is human progress), he proposed an assembly room. He suggested to the astounded natives that it is pleasant to stroll by the sea, and made a start on a marine parade; and he did even better by pulling off a great publicity coup, persuading the most famous Englishman of his time to bring his sickly young son – one day to be

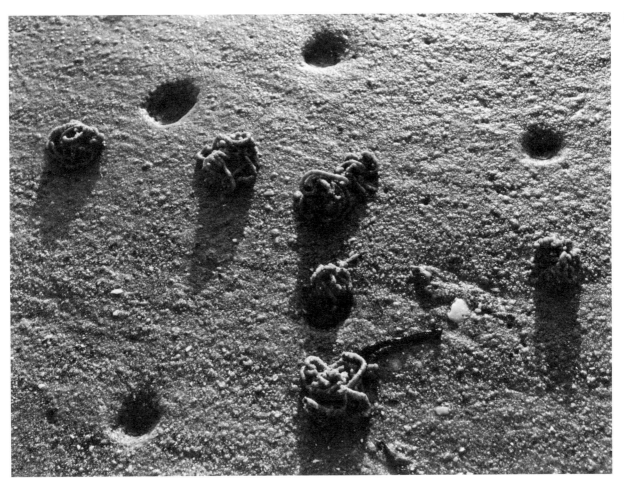

as famous as his father – to come to Lyme for the air and climate. If it was good enough for the Earl of Chatham and young William Pitt, it was very soon good enough for many others. Hollis, a much greater human being than either Pitt, performed this small miracle in only four brief years, since he died in 1774.

By 1800 the main industry of Lyme had become what he foresaw and what it has remained ever since – catering for those who come to the sea for pleasure. It took place as suddenly as this, between 1750 and 1780, in countless other small coastal towns all over Britain. The sea, its water, its air, the light and relief it gave landscapes, became the rage. Lyme received its most famous literary visitor in 1803 and 1804, Jane Austen with her family, and it is interesting to note the contrast between her judgments of the place and of its society. The latter received very low marks indeed from the mercilessly fastidious young woman, but she grows positively Words-worthian – if not downright brochuristical – when it comes to the natural setting. The lift, the *allegro* that takes place in *Persuasion* when the action moves to Lyme is completely typical of the people of its time. They had discovered what we are now taught to covet and love from infancy. I think there has been no nicer sea-change in social taste – even though in 1800 it was still reserved for the well-circumstanced – in man's history.

The hidden intention of that invisible turned-round siren installed on every beach was not at first perceived. For many decades sea-bathing remained what it had been to Jane Austen, a medicinal activity. Quite probably even fewer were actually bathing in the sea than at the very beginning of the mania, since along every promenade and front had sprung purpose-built interior (and warmed) sea-baths; and those who still braved Neptune direct did it from a wheeled cabin. But the Victorian spirit was dominant long before 1836, and it was that age which began to see the siren plainly – that is, the always implicit eroticism and sexuality of the beach.

No one saw it more clearly than the Reverend Francis Kilvert. He loathed 'the detestable custom of bathing in drawers' and twice – and delightedly – shocked public beaches by refusing to wear them. In 1873 he writes (one may take his 'ignorance' with all the salt in the English Channel) 'I had in my ignorance bathed naked . . . however some little boys who were looking on at the rude naked man appeared to be much interested in the spectacle, and the young ladies who were strolling near seemed to have no objection.' Two years later he was in the Isle of Wight during July.

The morning was blue and lovely with a warm sun and fresh breeze blowing from the sea and the Culver Downs. As I walked from Shanklin to Sandown along the cliff edge I stopped to watch some children bathing from the beach directly below. One beautiful girl stood entirely naked on the sand, and there as she half sat, half reclined sideways, leaning upon her elbow with her knees bent and her legs and feet partly drawn back and up, she was a model for a sculptor. There was the supple slender waist, the gentle dawn and tender swell of the bosom and the budding breasts, the graceful rounding of the delicately beautiful limbs and above all the soft and exquisite curves of the rosy dimpled bottom and broad white thigh. Her dark hair fell in thick masses on her white shoulders as she threw her head back and looked out to sea. She seemed a Venus Anadyomene fresh risen from the waves.

But Lolita-haunted Kilvert was a century ahead of his time in erotic honesty, and few others of his age would have admitted such thoughts, let alone committed them to paper. Yet they must have harried even the most timid and conventional. One may dress against other eyes, but not against the caress or shock of water on the most private skin. Decent Christian gentlemen and advisers of youth made such a thing of the manly cold bath because they feared terribly what might go on in warm ones. As late as 1882 the town council of Lyme was still threatening severe penalties on any male degenerate who dared venture within fifty yards of the ladies' cabins.

Our museum has a very revealing family album of 1886. It gives a delightfully vivid picture of what a seaside holiday was like in that year: prawning, mackereling, tennis, walking, fossiling, sandcastling, sketching, photographing, making fun of the locals . . . but not a single word about taking one's clothes off and swimming. The one other thing as conspicuously absent, to our own age's eyes, is symptomatic: despite the many – and evidently lively and attractive – young people of both sexes in the family, there is also not the faintest hint of any romantic attachment, even jocular.

'Sea-bathing tends to invigorate the whole nervous system,' pronounces *Modern Etiquette* in 1889. 'However, as an agent for promoting the softness and delicacy of the skin, and the bright hues of the complexion, it is inferior to the warm or tepid bath. It is better not to bathe in the sea until two hours after a meal, and the circulation should be promoted by friction and the aid of a good, brisk walk.' The lady authoress goes on, too, to warn against 'exposure to the ray of the sun in summer. It is very injurious to the skin, causing it to tan.' This latter reason was probably why the fashionable months for sea-bathing in Jane Austen's time were October and November; one had at all costs to preserve one's ability to blush. Nothing was more erotic to nineteenth-century man than a milky cheek turning pink.

But all this middle-class nonsense was doomed. The seaside jaunt had become more and more of a national habit, ever since Sir John Lubbock's Bank Holiday Act of 1871. Even *Modern Etiquette* had to confess, and approve, the fact that 'of late ladies have taken very much to rowing', while *Punch* had been hinting at the sexual charms of the seaside since at least 1864, when that trend-setting (and trend-mocking) Parisian, George du Maurier, joined its staff; and even then he was only taking up a line that the Late Georgian cartoonists like Rowlandson and Cruikshank had been as frank about as our own age. It was fully accepted by the 1890s. The jolly opportunities for studying the female form, for making the chance encounter, constitute the main theme of that charmingly illustrated series on popular European coasts by the French artist Mars – a series aimed quite as much at Anglo-Saxon audiences as at Gallic ones. All those hiked skirts and peeping lower calves, that wind-dishevelled hair, young Belle Epoque beauties in a disarray . . . from there it was only a short step – facilitated by Tommy Atkins's discovery of French popular art in his brief reliefs from the horror of the trenches – to the splendid vulgarities of the kind of postcard that George Orwell immortalized. I suspect we have still not fully recognized the debt sexual – and perhaps political – liberation owes to the seaside holiday.

It has now become the principal public pleasure area, closer and closer to the bedroom, of all advanced Western – and Eastern – societies. It is where one goes to spoil one's own naked body, to find sex and romance, to release; for an oblivion on all routines, fixed hours, formalities. It may become increasingly difficult to escape the world on the beaches of high summer; but even there workaday identity at least can be lost. All through every August I listen to the sound of children's voices that float up from the beach into my own seaside garden. They are within a quarter-tone of being screams of extreme terror; but remain screams of extreme pleasure. The subtle siren plies her trade, and meets very little resistance now.

Since the proximity of the sea melts so much in us, the island is doubly liberating. It is this that explains why indigenous small-island communities, at least in the long-discovered temperate zones, are on the whole rather dour and puritanical in their social ways and codes. They have to protect themselves against the other perennial temptation of the island: to drop the necessary inhibitions of mainland society. Islands are also secret places, where the unconscious grows conscious, where possibilities mushroom, where imagination never rests. All isolation, as the cold bath merchants also knew, is erotic. Crusoes, unless their natures run that way, do not really hope for *Man* Fridays; and islands pour a stronger wine of

forgetfulness of all that lies beyond the horizon than any other places. 'Back there' becomes a dream, more a hypothesis than a reality; and many of its rituals and behaviours can seem very rapidly to be no more than devices to keep the hell of the stale, sealess, teeming suburb and city tolerable.

The puritans, from Homer on, have always suspected islands, and wished their addicts the fate allotted Odysseus and his men. William Golding repeated the ancient warning in *The Lord of the Flies*: such literal isolation will breed swine – self-destruction, whether it comes through lotus-eating or loss of mainland law. I think it is significant too that the most self-revealing novel Thomas Hardy ever wrote, *The Well-Beloved*, is set on the quasi-island of Portland. It is a story full of incest, of repressed eroticism, of narcissistic guilt in its tortured author. He makes much play between the pagan and the Christian view of life, the illicit and the licit; and makes it very clear that the illicit inhabits the old Portland (and his own complicated psyche) precisely because of its detachment from the mainland, both physically and psychologically.

I have always thought of my own novels as islands, or as islanded. I remember being forcibly struck, on my very first visit to the Scillies, by the structural and emotional correspondences between visiting the different islands and any fictional text: the alternation of duller passages, 'continuity' in the jargon of the cinema, and the separate island quality of other key events and confrontations – an insight, the notion of islands in the sea of story, that I could not forsake now even if I tried. This capacity to enisle is one I always look for in other novelists; or perhaps I should say that none I admire lacks it. It is a capacity that lies quite literally at the heart of what has often been called the first modern novel, Defoe's *Robinson Crusoe*; and it lies equally at the heart of the very first novel of all. The island remains where the magic (one's arrival at some truth or development one could not have logically predicted or expected) takes place; and it rises strangely, out of nothingness, out of the onward dogwatches, mere journeying transit, in the writing.

The Scillies have a Victorian novel partly devoted to them. This is Sir Walter Besant's *Armorel of Lyonesse*, first published in 1890, and not possible much before that date, since it shows the 'new' young woman of the time in all her brave and earnest glory. One must in art derive the species from the Pre-Raphaelites, but the vogue for her sturdy, glowing cheeks, her impetuous frankness, her comparative emancipation, probably began with another novel ten years earlier than Besant's: Sabine Baring-Gould's *Mehalah*. The similarly eponymous Mehalah is also an expert oarswoman and girl-mistress of an island, though of a very different kind from Armorel's Samson. The two books, like their two once famous and now forgotten authors, make an interesting contrast. Baring-Gould, an embittered High Anglican clergyman, took the puritan view of islands, which allowed him to write a much darker and finer story. Swinburne noted its powerful echoes of *Wuthering Heights*, another essentially 'island' book, for all its geographical setting; and I have sung *Mehalah*'s praises elsewhere myself.

Besant was a better man than writer: a lifelong champion of the under-privileged and a patron saint to all American and British authors, since he was the first to campaign consistently and effectively for their legal and commercial rights. The London scenes in *Armorel* are rather too full of stock characters and stock melodrama to make the book any but a very minor work of its period. Yet for thirteen chapters in the beginning, when Besant describes the life of his young heroine on the now uninhabited island of Samson, he does achieve something better, a strange sea-idyll, or island pastoral; he even achieves a genuine echo of a much more ancient girl on an island, and not only because that far more famous idyll, which I shall come to, also begins by the saving of a selfish, handsome stranger from drowning.

We may smile today at the passionately idealistic tomboy Armorel, who is several light-years removed from our own notions, both political and physical, of the aware and attractive young woman; but something of the honesty and independence that islands bestow, and of their ancient magic, does glow through these pages . . . one doesn't forget Armorel, or her spirit, indeed may even regret them, since the conditions of solitude and of self-dependence where such character can emerge are now gone from the world. But they are, still today, a little less lost in the Scillies than anywhere else in Britain south of the northern isles.

The word isolation comes, through Italian and French, from the Latin for island, *insula*; from which we more directly derive *insulation*. Thus from the same mother-word we have both the active and the passive, the painful and the protected aspects of being cut off. The ancients rather prettily derived *insula* from *in salo*, 'in the salt sea'; but it comes from a primordial and widespread root that also surfaces as the Greek *nesos* and the Irish *innis*.

Our own quite disconnected *island* comes from an Old English word that had no *s* – it crept in, unpronounced, only by scribal confusion with the medieval French *isle*. Yet there is a kind of justice in that surreptitious *s*. For *s* and *l* are great water-consonants, the sound of the sea and the sound of the sea on land, surging and washing, lashing and lapping; *insula*, isolation, soleness, solitude. 'Soul', in the dark backward and abysm of linguistic time, is another universal European word, meaning transient, and sounds much more maritime and soughing in its Old English form, *sawol*, to which the Greek *aiolos*, fleeting, shifting, as changeable as the wind-god Aeolus, is connected. And what more natural than that ancient man should gain his sharpest sense of both physical and metaphysical loneliness ('all-oneness', nothing but one soul) at sea, and associate the concept to its sound – perhaps above all on the most formidable of the seas he knew, and named after At*las*?

That the sea does speak these consonants *s* and *l* is most clear in calmer and more Mediterranean weather, and especially above shingle beaches at night. The impact of a small wave is susurrant and palatalized, high-pitched; its lazy withdrawal, lower and gargling; and one can very often, on a long beach, hear both sounds distinctly yet simultaneously. It is very familiar to me, since I spend a last minute or two, almost every night of my life, listening to what the always proleptic sea is saying below my house. I do not think it is a coincidence that the most ancient word we English (and our North European cousins) use for travelling on the sea also contains those two sounds. 'Sail' is well called. That *s* and *l* wash back in the commonest, the 'salt' that appears in one form or another in every European and Slavonic language, and the rarer – for instance, in that quintessential oceanic bird, the solan (from the Old Norse *sula*) or gannet. But above all I like it in our equally ancient name for the marine creature that most resembles man – and which gave birth, by its appearance and its voice and its favourite habitat, to many of the mysteries and confusions I shall discuss in a minute. Creation played a joke when it made seals prefer the places, like the Western Rocks of the Scillies, where sailors are most often wrecked.

Another familiar collocation is between tree-sound and sea-sound. If I do my nocturnal listening on the landward side of an old Scots pine in my garden, there are conditions of wind when one cannot distinguish wave-

sound from foliage-sound: they blend perfectly. The ancient forests of Europe were a very near analogy to the sea in their dangers and loneliness, and we have always used maritime metaphors to describe their vastness . . . an endless sea of trees, and so on. The Latin word for forest was *silva*, related to the Greek *ule*, of the same meaning; which is in turn connected to *ulao*, to bay or bark or ululate. Wolves. It is this savage (from the Romanic *salvaticus*, wooded) and primeval hidden fang in both maritime and forest solitude that I am steering towards.

No one is quite sure where the word Scilly comes from. R. L. Bowley suggests it is Celtic, Old Cornish, of unknown meaning. I will quote Geoffrey Grigson in full:

> Several times I have referred to the islands, not as the 'Isles of Scilly', but as 'Scilly' – which is still common usage. 'Siluram insulam' is how the islands were mentioned – if, indeed the Siluram is Scilly – by Solinus in A.D. 240; Sulpicius Severus about A.D. 400 wrote of 'Sylinancim', again in the singular; and so it goes on – 'Sully', 'Sullya' – down through the Middle Ages to the modern 'Scilly'. What has been for so long a convenient collective name may – *may* – be indicative of the ancient unity of most of the islands. The *y* of 'Scilly' is probably the Old Norse *ey*, 'island' (though the root syllable is ancient and obscure).

I will add that the name given in the Icelandic *Heimskringla* (about 1230) is Syllingar; and that the Roman name for the Bishop and Clerks off the south-west Pembrokeshire foot of Wales, similarly placed to the Scillies, was Silimnum. It is possible that there is an association with the Silures, the southern Welsh clan whose capital was Caerleon.

Now I am going to rush in where etymological angels fear to tread. Something ominous haunts this *skl* or *scl* conjunction of consonantal sound; a plosive crash between the surge and the lap. If *sl* is solitude at sea, *skl* is death there. *Sk* on its own (the hypothetical Indo-European root *sek-*) is unmistakably associated with tearing and cutting; another ubiquitous root *se-* means 'without' or 'apart' – loss and separation. There is in fact an Old Norse word, *skil*, that might sit with the *ey*, or island. We get our modern 'skill' from it, but the noun came from the verb *skilja*, which has closely related forms in the other Germanic languages: they all mean 'to separate', thence 'to distinguish or differentiate' . . . the basis of all skills. Schools of whales, for instance, have at first sight etymologically nothing to do with schools of learning. We spell the two words the same, but the whale *school* comes through Dutch from a Germanic word related to *skilja* – that is, it is a division or section of whales.

We derive 'school' in the other sense through Latin from the Greek *schole*. But the original meaning of that word, as countless reluctant young Greek scholars have no doubt wistfully noted, had nothing to do with boring old work at all. It meant the very contrary: leisure, idleness, time out from real labour . . . in which one could discuss the world. Then *schole* became the place where such discussing took place and so, creeping like a snail, to the sense we now give it. But it too began merely as something divided, cut off from normal activity.

The 'Sully' and 'Sullya' spellings – Sorlingues in French – that Grigson mentions might then perhaps mean something like 'separate island' or even conceivably 'island like a school of whales'. Whale Island is a common enough name elsewhere, and the Scillies can be whalebacked at a distance, from sea-level. But the sense would more probably be active and sinister – the 'separating islands', cutting of the thread of life . . . the dangerous islands. The fact that many of the Scillies' other sobriquets – Islands of the Blest, the Fortunate Isles and so on – suggest the very opposite is no counter-argument. One of man's oldest fetishes, perhaps our last fully retained feature (especially in the higher reaches and watergates of society) of palaeolithic psychology, is that of calling evil by a good name. I can no more credit that living ancient sailors found the Scillies 'fortunate' than that the Greeks believed the Eumenides were indeed 'kindly intentioned'. Such terms are propitiatory, not honest.

Another not implausible Norse origin lies in the word *skiallr*, which became *sciell* or *scyl* in Old English. We derive the modern 'shrill' from it, but it had a strong association with noisy water – resounding, roaring, restless. There are three villages called Shillingford in England, two in Devon, and one in Oxfordshire: among their medieval spellings are Sullingford, Syllingesford and Schillingforde, which do bring us close to the Silura, Sylinancis, Sullya and Syllingar of the oldest records of the modern name. But since we are with Silura back in the third century A.D., well before the Anglo-Saxon or Viking arrival in the south-west, it seems to me that only the terminal syllable in Scilly can be reasonably guessed to be Scandinavian; and for the first I have a suspicion we should set sail for much further south.

Where did the sibilant sirens (Greek *Seirenes*) reputedly feel most at home in antiquity? On rocks near Cape Pelorus in Sicily, the Greek *Sikelia* – again that *skl*. The oldest name we know for this endlessly contested island is Sicania, but its original Sican inhabitants were invaded from the north, like Greece and Asia Minor, about 1100 B.C. The invaders, no doubt as ruthless and determined as the Dorians who destroyed Mycenean Greece, were the Sicels, originally (like the Dorians) from the east side of the Adriatic, another dangerous and reef-ridden coast.

We can fairly safely guess that the sirens were in fact a colony of *Monachus monachus*, the now rare but once common Monk Seal, for reasons that will appear. Cape Pelorus is the modern Cape Faro at the northern end of the Straits of Messina and in Roman times there was a town on the mainland directly opposite called Scylla. It was named after the most dreaded sea-monster of antiquity, the Greek Skylla.

Skylla has an exceedingly muddled ancestry in mythology. Her parents vary. In one version they are the sea-god Phorcys (usually allied to Latin *orcus*, a whale, but *phoki* is the Greek for seal) and Ceto, or Hecate Crataeis, 'powerful witch', a sinister dog-headed (like a seal) divinity related to the Egyptian jackal-god and harbinger of death, Anubis. In another version the parents are Typhon and Echidne, who was a woman above and a spotted (like a Monk Seal) serpent below, and who also gave birth to a formidable brood of other freaks. One was Cerberus, the three-headed dog-guardian of the gates of hell. Two others were Hydra, the many-headed water-serpent, and Orthrus, a two-headed watchdog whose first master was Atlas and whose killing at the gates to the Atlantic was the tenth of Heracles' tasks – and the one that most strongly echoes the even older Sumerian legend of Gilgamesh's voyage to the land of the dead. I have a little bronze statuette of Gilgamesh, made some 3,000 years ago, on a shelf beside me: and there he is, still smiling proudly as he wrings the necks of the two beasts, death by water, Orthrus, that have sprung for his throat. We remember Orthrus better as Sirius, the Dog-star; which should be the Seal-star. *Seiren*, a siren, *Seirios*, the dog-star, *seira*, a chain or noose. Hounds of death, many-headedness, the association with the sea: out of the water seals like to lie tightly packed.

Skylla's father Typhon was hardly more attractive. He was a huge fire-belching thug, with arms six hundred miles across, festooned once again with serpents' heads and tails. He scared the immortal daylight out of all the gods and once even captured Zeus and imprisoned him in a cave guarded by his sister Delphyne (dolphin). But Zeus escaped and eventually did for the monster by throwing Mount Etna on top of him – which is why it still smokes and rumbles. This neat ending is pure fiction, since Typhon

remains very much with us. He was always a great traveller. There is an Urdu word *tufan*, even a Chinese *tai fung*. *Tai fung* means big wind, or typhoon.

In spite of this dubious parentage all the variants agree that Skylla began life as an attractive young woman – too attractive for her own good. Again there are different versions of what happened. In the seemingly oldest she had an affaire with the Cretan sea-emperor Minos (or his divine corollary Glaucus), which displeased his wife Pasiphaë (alias Amphitrite, the mermaid goddess). In another version the illicit affaire was with Poseidon, and in yet another the offended third party was the island-witch Circe, who happened to be in love herself with Minos-Glaucus. Circe decided to ditch her rival for good, so poisoned the fountain where Skylla bathed. Everything below Skylla's waist was changed into barking dogs (Greek *skyli*, a dog); above, she found she had six heads with three sets of teeth each. In understandable despair she threw herself into the sea between Italy and Sicily and made her home in as dangerous a set of reefs as ancient sailors knew – and became linked in proverb with the nearby whirlpool of Charybdis. She also had her revenge on Minos both for not standing by her and for killing her father, since he too met his end in Sicily.

Skylla and Sikelia, Scylla and Sicily, seem then to stem from a very ancient association (not of course confined to Sicily, but the Straits of Messina were a vital short cut for ancient sailors) between death or great danger at sea and baying hounds with bodies like women – the seals. 'Seal' is a shortened form of the older word, still revealed in the Scottish *selch*. It was *selich* in North English, and a Germanic word *selchaz* (the *ch* as in 'loch') 'of unknown origins behind it. As I said, the Greek for seal is *phoki*. But by another oddity, if we bear in mind the association between seals and dogs, teeth and reefs, there is a Greek word *selachos* that has very much to do with the ravenous sea. It meant shark, or dogfish.

It so happens that the uniquely Mediterranean (and Black Sea) Monk Seal is far more sedentary and faithful to its chosen home than the other, and pelagic, European species; and perhaps one good came from all this – that its now threatened extinction was so long delayed because of the superstitious fear the poor animal provoked in the ancients. Though clearly propitiatory in intent, a suspiciously high proportion of the traditional individual names of the sirens do contain a howling *l*: Molpe ('melody'), Aglaope ('beautiful'), Thelxiope ('enchanting'), Leucosia ('clear-voiced'), Ligeia ('piercing, sweet-toned'). Many less polite ancient accounts state that Skylla's voice had a strange whimpering hound-like quality, as indeed a distant seal-colony might sound at night.

The Grey Seal's voice is described thus in the *Field Guide to European*

Mammals: 'voice loud, long-drawn-out, two or three-syllabled (hah-ee, hou)'. The Monk Seal makes 'a repetition of o- and ah- sounds'. The three strict species of Monk Seals are all now extremely rare, but related Antarctic Monachinids are loquacious. The Ross Seal coos and trills. Weddell Seals are 'very vocal' during the breeding season and are often heard calling beneath the ice, a cry that begins high-pitched and then descends. The Leopard Seal is the same. On still nights in sailing days, these mysterious sounds may well have been audible. Geoffrey Grigson speaks of watching the Grey Seals on White Island in the Scillies and listening to them 'moaning like children in pain'. And he tells the story of a mid-Victorian blacksmith, left alone on Rosevear among the Western Rocks during the building of the Bishop's Rock lighthouse, who swore blind the next day that he had heard singing and music in the night. Even the prosaic Nansen was stopped one day during his epic Arctic traverse with Johansen. They heard a mysterious sound 'like a goat's horn' floating from some huge distance over the icy wastes, where no seal was visible.

There is a stranger ululatory link still, with that other sea of the great forest, since another part of the Skylla legend links her with the moon-goddess, who traditionally assumed the shape and voice of an owl in her death-phase. 'Owl' and 'howl' both derive from some common word behind their Germanic and Greek roots, and one of the most universal words of all is 'wolf' – the Old Norse *ulfr*, the Old Slavonic *vulku*, the Greek *lykos* are all connected. That ubiquity is shared by the word man first applied to the wolf domesticated, *hound* in its English form. Baying and howling are, of course, death omens in all folklores.

That there was a particularly sinister aura about the seal can be confirmed by the distinction the ancients made between malevolent and benevolent sea-creatures. The first were Phorcids, the latter Nereids – in natural history terms, the animals found round dangerous reefs, the seals, and those met safely out at sea, porpoises and dolphins. That is why there has always been a strong dislike among fishermen and sailors of killing the latter: and why there is such an age-old dislike of taking dogs on board ship. The equally ancient hatred of whistling and of women, and especially of the two in conjunction, is associated. Many of the taboo animals for the sailor – the dog, the hare, the rabbit, the pig, the rat – are the same for those who risk their lives underground, the miners. We may think such superstitions are dead now. But as recently as 1959, a crew of fifty men got off a cargo-ship at Liverpool and refused to continue their voyage. They had discovered in the course of a bad transatlantic trip that the master had smuggled aboard a bird in a cage for a German zoo: it was a live albatross. It died the day the ship docked.

Gull is a Celtic word (*guilan* in Old Cornish), but again linked with a universal Indo-European root that carries a general signification of swallowing (gulp, gulf, gullet, gulch and so on). And if one traces back the complex-historied word albatross far enough one ends with a shared Greek, Arabic and Hebrew word meaning a bucket . . . a swallower of water and all who sail on it. Sea-gulls, with their need to raise their necks and open their beaks both to gulp and to call are eternal reminders of what the sea can do, and that is why sailors have from the very beginning of time been careful not to injure or insult them; and why, before 1800, the washed-up bodies of the drowned were far more often buried below the tideline than in consecrated ground. What the sea swallows, it likes to hold for ever. This respect for the sea and its emblematic creatures antedates our modern recognition of gulls as useful scavengers by several millennia. An old dialect word for the bird, *scull*, is another candidate for the naming of the Scillies; and returns us to that same sinister conjunction of sound.

Lemprière mentions other names that exhibit it. Scylacium was a sea-coast city of Calabria, nicknamed by Virgil *navifragum*, shipwrecking. Scylax is an obscure pre-Herodotean geographer, of the time of Darius, who evidently wrote an account of a sea-voyage and 'an enumeration of twenty important islands'; an odyssey, in short. Cape Skillaion is a nasty corner of the Peloponnesus – I have sailed round it many times and can vouch for its ugly seas in bad weather. The professional descendants of Scyllias are very common in the Scillies today: he was the first wreck-diver known to man, also mentioned in Herodotus.

Askalon was a sea-port, near the modern Ashqelon of Israel. It dates back at least to 3000 B.C., and belonged to the historically traduced Philistines (a

comparatively civilized people for their time), forming part of the Phoenician-Tyrian sea-empire. That is, it was among the great explorer-sailor cities of the ancient world. The presiding god of the sea-going Philistines was Dagon, a merman, related to the Nereus-Proteus (old man of the sea, first man) of Greek mythology. The chief temple at Askalon was that of a corresponding female deity, Atargatis (also worshipped in Cyprus and Crete), who was a woman with a swimming tail . . . a seal-siren, ancestress of Amphitrite and Aphrodite Anadyomene, Venus rising from the foam.

The initial A of Askalon carries a negative sense, and classical scholars usually interpret it as meaning 'unploughed' or 'unharrowed'. I think myself that the name may be propitiatory, and carry a significance of 'unshipwrecking', beneficent to ships. The name of the Greek god of medicine, Asclepius, is formed in the same negative way: it means 'unrending' – not causing, or preventing, pain.

Askalon is remembered in English because it grew a famous sea-onion, one that has pleased gourmets all through history. The 'onion of Askalon' is the scallion, or shallot, of every good vegetable garden. It is not only Anglo-Saxons who cannot abide garlic: devils and witches hate it equally. All onions have strong and universal protective powers in folklore. One genus of the family, the squill, or *Scilla*, even bears the reef-monster's name. The medicinal use of the sea-squill proper, *Urginea maritima*, dates from the very earliest times, and its chief producer, then as now, has always been Sicily. Squill bulbs were once carried by travellers on both land and sea as amulets; and hung round sheep's necks as deterrents to wolves. Nothing defanged destiny more surely than these potent plants.

The *skl* evidence of any Greek dictionary (let me just murmur skeleton) is even more telltale of maritime connection and gruesome disconnection . . . the sea-coast as hideous archosaur, ultimate wolf, ripper of the sailor from his flesh and his life. Old Norse *skilja*, if you remember in all this killing (mystery word, origin uncertain, but consider cleaving and cliffs) and culling, bears a primary meaning of 'to separate'. Sea and separation skulks in the sound both north and south; in the bygone English scacol (a sea-cape), scald (a shoal), scalp (a tidal rock), skull and scull (which cleaves water), scylfe (a crag); in the Latin *scalpere* (to cut), *scelus* (evil deed), *secula or sicilis* (a sickle), *silex-silicis* (flint) . . . with which last man first cut, and which both Virgil and Lucretius used as a metonym for 'sea-cliff'.

Sea-cliff. There is no escaping that sound.

As regards where the word Scilly came from this is all the purest speculation, or looking out to sea; but I incline to think we should not take universal onomatopoeias too lightly, and that there is rather more than Old

Cornish or even Old Norse in the name. It bears the ghost of a sound that from the very beginning of human speech has conveyed cutting, rushing, splitting, stripping, tearing flesh from bone . . . toothed dogs and sea-demons despoiling all. And if you ever catch the Scillies in a big sou'-wester, you will wonder no more: Skylla still roars and screams and whimpers from every rock and cliff and shore.

The first novel in world literature is woven of islands and the sea, and of solitude and sexuality, which is why it has had a greater influence on subsequent story-telling, both thematically and technically, than any other single book in human history. It also first demonstrates (with a complexity and subtlety that still escape all probability) the value for the form of the archipelagic structure I spoke of earlier. A novelist can no more afford not to be steeped in it than a Christian in the Bible, a philosopher in Plato, or a socialist in Marx. It is the *sine qua non* of all serious study or practice of fiction.

I am one of the heretics who believe the *Odyssey* must have been written by a woman. The heresy is not new among authors. Samuel Butler believed it, and produced some convincing circumstantial evidence; and so does Robert Graves, in our own time. Whoever did write it seems markedly more knowledgeable about domestic matters, the running of a large household, than about nautical ones. The one bit of showing-off in the latter field – the description of the boat Odysseus builds to escape from Calypso's island – is shipyard stuff, not sea-going expertise; and in the very first pages nothing is more striking than the loving detail bestowed on the *provisions* for Telemachus' voyage and the total absence of such detail when it comes to the craft itself. Throughout history it has been man who worships and polishes the vehicle, and woman who packs the suitcases.

Transparently also the writer is obsessed by all the things – especially young female things – that keep husbands away from their proper place at home. There is that repeated, vivid eye for the interior decor and life of the palaces of Nestor, Menelaus and Alcinous – how guests are received, how they are bathed, how dressed, how fed, even how the laundry is done; that ubiquitous sympathy for the feminine ego, from the glittering *grande-dame* entry given Helen of Troy to Calypso's sadness; the love of describing clothes and jewellery; the sympathy shown older women, the flagrant greater interest, in the Land of the Dead, taken in the female ghosts . . .

Butler, who was no Kilvert, decided the authoress was hiding behind the nicest (morally) of the waylaying island girls, Nausicaa. If the writer must be hidden behind a character, I should plump myself for Penelope, or rather, for the theory that Scheherezade was not the first woman to know that letting a man hear all that he imagines is one very good way to put him under your spell. Who would want the cold, salt reality after such a telling?

As in so many other matters, and for obvious physical and social reasons, it seems probable that if man went out and brought home the raw material, it was always woman who cherished, 'cooked' and wrought it. With men, it was always the challenge of getting; with women, the elaboration of the got. We know that women tend to be the main 'carriers' of folk-song and folklore among primitive peoples. Men must perforce have had a closer knowledge of external reality, however superstitious they were; and women a closer knowledge of the internal imagination – of the store-room of the reported image, not the directly apprehended one. Weaving and embroidery lie at the heart of all story-telling, as they do at the root of all decoration. The Greeks knew it. Their very word for a recited epic, rhapsody, means simply 'stitched song'. Plaiting the real and the imagined defines all art; and I think it no coincidence that (like Circe and several other women in the book) Penelope took to weaving as both her pastime in her long wait and her excuse for her fidelity.

The more one looks at the internal evidence, the more convincing it is. Who steals the very first chapter of the great story? For whom is most sympathy evoked? Not for the absent Odysseus, but for his abandoned wife and all her domestic problems, most strikingly that of a son well on his way to becoming another Orestes. And what is the emotional climax? The night

of reunion in Book XXIII, when even dawn is delayed in awe . . . on the command of the divinity who has finally brought patient wife and wandering husband together again – another woman, the wisdom-goddess Athene. (The actual final book, XXIV, is a mere tidying up of loose ends.) No man who has ever risked or provoked the shipwreck of a marriage by his own selfishness has ever doubted the profound affirmation of *female* wisdom in that climactic passage; or for that matter had to wonder why the ancients personified wisdom as a woman. Even more significantly Athene is a pre-Greek deity. She was the protectress of palaces in the Mycenean Age where the story is set; and also the goddess of arts and crafts . . . a women's goddess, if ever was.

We know that behind the Homeric legend of the Trojan War lay a very real conflict in the last centuries of the second millennium for trading power and land to settle. It was between a loose confederation of Mycenean pirate-kings and the holders of the gate, the Bosporus, to the coveted Black Sea. We also know that 'Homer' was writing several centuries after these events and by no means (although one may argue over how conscious the irony is) with undivided admiration for the Mycenean part in them. Few of the male heroes – human or divine – are very attractive, or allowed to be happy, and especially when they are away from home. Zeus and Poseidon move only to punish; moving men invite punishment. Penelope's suitors are continually being told to go home; and refusing to do so, duly meet their end in a blood-bath, a mirror-image of the agonies that Odysseus, the sole survivor of his Ithacan squadron, has been through on his own travels.

What bouquets there are for men go to those who have either stayed or resettled at home: Menelaus, Nestor, Alcinous, the prince-shepherd Eumaeus, old man Laertes. The enigma, of course, is Odysseus himself. In one light he is the least attractive character of them all, with his compulsive lying, his suspiciousness, his infidelity, his vindictive anger – but then his very name means 'the one with enemies', the 'victim', or in simpler modern terms, the paranoiac. In this aspect it is difficult not to see him inside a much more recent myth, against a background of tiny islanded townships in that other wild ocean of the American West; and most certainly not there with the face of the noble sheriff, but much more with that of a Lee Marvin or a Jack Palance . . . the unscrupulous, pathological killer. Dryden found two perfect adjectives for this aspect of him in his translation of Virgil's *Aeneid*: dire and insatiate Odysseus.

If that was all there was to him . . . but there remains his courage, his onwardness, his questing, his surviving, his shrewdness, his humour, his fallibility; his quintessential maleness, with all the faults and virtues, as closewoven as the shroud-cloth on his wife's loom, of that biological

conditioning. One has only to compare him with the heroes of the other Ancient Greek travel-sagas, Heracles, Perseus, Bellerophon, Jason of the *Argo*. They are myth-puppets from the nursery of the imagination. Odysseus is real, and human, however mythical and supernatural the circumstances in which he finds himself. If his vice is that he cannot stay out of the game, his own cheating – mainly done merely to survive – is petty compared to that of the other players, the gods who control the way things are; which makes him universal man, and justifiably paranoiac.

The *Odyssey* is fundamentally an analysis of the mechanism and the justice of this paranoia. Odysseus' greatest and most implacable enemy, Poseidon, is the god of the great medium of his temptation, the sea. His return from Troy is therefore both a penance for past sins and a running demonstration of why they came about. Again and again he or his men have their agony prolonged because of their own greed or pointless aggression. His one virtue is his longing to return home, to find Penelope – that is, wisdom; but (despite the night of reunion) Homer puts a great question mark over this. The savage massacre of the suitors, the hanging of the corrupted maids, the mutilation ('with a sharp knife they sliced his nose and ears off, they ripped away his privy parts as raw meat for the dogs, and in their fury they lopped off his hands and feet') of the shepherd Melanthius, all suggest the paranoia remains. And there is a famous thread left lying loose at the end, the fact that Odysseus knows from Tiresias that he must make one more voyage still: 'till you reach a people who know nothing of the sea and never use salt with their food'. That is, he must travel for ever, on this planet, in search of an unattainable, his own landlocked peace. The sea, the invitation to the unknown, will remain his unassuaged demon.

The *Odyssey* has always strongly reminded me of one side of a later literature, also written in a time of struggle for power and search for *Lebensraum*, of quasi-ritualized aggression, of brute male greed for prestige and property, and all in the context of another line of ambitious sea-pirates. The period following the Norman Conquest was also that in which the first women writers of our era began to use story-telling – based, as with Homer, on material from long before their own time – gently to suggest better ends in life to their menfolk. The key vehicle for this new world-view was once again the story of wandering adventure, though their protagonists' sea was more often the forest than the literal one. But true sea-voyages and islands are by no means absent, and I have already suggested that the parallel between the old, vast, mystery- and wolf-filled forests of Europe and the sea was very strong.

What is significant is that writers like Marie de France and Christine de Pisan chose to send out so many latterday knightly Ulysses on their voyages

of self-discovery; and their very frequent final demonstration that true wisdom always lay at home, or quite certainly not in the overt original purpose of the journey. It is very instructive to read the *Odyssey* and Marie de France's stories side by side: it is not just the central similarity of attitude to the quest theme, but the little touches of humour, the psychological accuracy underlying the delight in the fabulous (the ability to make fabulous beings behave humanly), the obsession with domestic behaviour and domestic objects, the preponderant role played by the relationships between men and women . . . a shared set of sensibilities and pre-occupations that we know, in the latter case, did not belong to a man.

Even if one must take the orthodox scholarly view, and make Homer the male bard that tradition has always maintained, it seems to me certain that he was composing quite as much for a feminine audience as a masculine one, and from an essentially feminist point of view: that is, a civilizing one, and using very much the same techniques as those early medieval writers. Scholars have delighted in seeing the *Iliad* and the *Odyssey* as anthropological crossword puzzles all of whose clues lead to solutions in obscure religious symbolisms: Penelope becomes the centre of a duck-goddess cult, Odysseus a sacrificial king, and so on. Of course that is a part of it. But I think no one who has wandered round the palace-complexes of Knossos or Mycenae can believe that even in that time (far before Homer's) there cannot have been, behind the picturesque sacred groves and golden boughs, ordinary men and women with practical social problems; and quite sophisticated enough mentally (if their artefacts are anything to go by) to distinguish at least some of their contemporary relevance from their mythical representation.

Archaeology has time and time again proved that the Homeric descriptions of artefacts and techniques were very far from mythical; and the continuous and highly realistic central theme running behind all Odysseus' ordeals and adventures is the predicament of the wife left to rule in her husband's absence. This must have been a very familiar one in an age of universal piracy. No theme is more often repeated in the *Odyssey* than the upset to the economy of Odysseus' palace – that is, to any island-state without a firm hand in control. There was, of course, in the chronology of the story, a very recent example of just this problem not far from Ithaca – the adultery of the queen of Mycenae with Aegistheus, their murder of her husband Agamemnon on his return from Troy, their own murder in turn by Orestes . . . forced marriages, usurpations of power, internecine and family strife, endless petty war. And why? Very largely because of male stupidity and arrogance, that inability to be satisfied with what one has, that perpetual lust to amass more, possess more, score more – and that last in a very modern sense, since female slaves were a not unimportant part of the Mycenean marauder's hoped-for booty.

The sea was the road to this lust; and also an escape from wisdom and the wives at home. Odysseus exists in a web of women of all sorts, both the wise and the wicked; and again and again he is saved from the wicked ones by Penelope and Athene. The vivacious, games-playing and deliciously polymorphic goddess can be read as a spirit version of Penelope – an Ariel to her Prospero – since she is obviously half in love with Odysseus herself, reproachful of his other women, yet not for a moment jealous of his wife. She even makes her more beautiful than she already is, using a divine cosmetic to make her skin 'whiter than ivory'. She is a wish-fulfilment of the woman left at home, in other words; or of the writer and his or her audience.

The sea and its islands thus become the domain of what cannot be controlled by wisdom and reason; the laboratory where the guinea-pig Odysseus must run through the mazes; where the great ally of reason, the conscious, gives way to the rule of the unconscious and the libido, that eternal and oceanic unsettler of domestic peace and established order. Since it is Odysseus' own unconscious that drives him on, its sea-domain is peopled by women. There is perhaps no more brilliant antedating of Freud in all ancient literature than that meeting Odysseus has, on the dark shores of the River of Ocean at the end of his furthest voyage, with his own dead mother Anticleia. 'As my mother spoke, there came to me out of the confusion in my heart the one desire, to embrace her spirit, dead though she was. Thrice in my eagerness to clasp her to me, I started forward with my arms out-stretched. Thrice, like a shadow or a dream, she slipped through my arms and left me harrowed by an even sharper pain.' In that brief image lies the genesis of all art; the pursuit of the irrecoverable, what the object-relations analysts now call symbolic repair.

There was a celebrated dispute in the last century over Cinderella's glass slippers – the *pantoufles de verre* of Charles Perrault's original seventeenth-century French text. All the most learned authorities, including Littré, declared that *verre* must have been a misprint for *vair*, or ermine. Glass slippers were not logical, what is not logical must be an error; which in a fairy story is the greatest illogic of them all, as Anatole France finally pointed out to the very learned gentlemen. Something like a glass and fur dispute has always surrounded the itinerary of Odysseus. As it stands it is so confusing that it is a less kind reason to believe no man could have written the account of it. But I suspect that even for its own time it was already at least half a voyage of the imagination, although undoubtedly fuelled by tales from Phoenician, Tyrian and Greek sailors.

Apollonius of Rhodes, who was a librarian and scholar as well as an odyssey-writer himself, suggested in the third century B.C. that the *Odyssey* is in fact a journey around the coasts of our old friend Sicily – one long thrust-and-parry with Scylla and Calypso, the island as murderess, the island as seductress. One of the various islands named after Calypso that dot every classical atlas does in fact lie in the arch of Italy's mainland foot, near the city and shipwrecking gulf of Scylacium (modern Squillace) mentioned earlier; and ominously off a Cape Lacinia (Greek *lakis*, a tearing or rending). If Homer, male or female, did live (as is generally accepted) in Greek Asia Minor, then Sicily is sufficiently far off to have been a suitable setting.

Odysseus' wanderings begin, once the Trojan war is finished, with a characteristic raid on a sea-coast city in Thrace, Ismarus. The crews of his twelve ships commit a wholesale slaughter of the men, take the women and sack the town. Odysseus argues for a swift withdrawal, but his sailors drink and feast, which allows the inhabitants further inland to counter-attack. Odysseus retreats to sea after a futile loss of seventy men. A gale drives him to Libya and the lotus-eaters – the island as *dolce far niente*, oblivion. The three men sent as a reconnaissance party fall for the lotus and have to be forced back on board. From there the doomed squadron sails towards Sicily, and encounters the one-eyed sea-killer Cyclops. Odysseus gets away by blinding the giant, who hurls huge rocks in his fleet's general direction – a reference perhaps to Etna, or the island of Strongyle (Stromboli) just north of Sicily. This is what infuriates Poseidon so much against Odysseus. It is true the giant had torn the limbs off several of his men and eaten them for supper, but once again Odysseus (his maternal grandfather was the notorious cattle-thief Autolycus, or 'wolf') was the prime aggressor.

From there he sails north and comes to the island of Aeolus, Guardian of the Winds, whose six sons have married his six daughters, in the age-old incestuous way of island people. There Odysseus and his men are well treated, but when they leave he is given a bag of winds, with strict instructions never to untie it, since the one wind left outside and free to blow is the Western, which will take them back home to Ithaca. And home they almost come – Ithaca is even in sight – when the crew decide (male greed again) that the mysterious bag must contain treasure Odysseus is keeping to himself. They open it secretly – and are promptly blown all the way back to Aeolus. This time the wind-king is less kind. He will not help them any more.

The little fleet's next landfall is the island of the Laestrygonians. Odysseus begins to learn his lesson; he moors his own ship where he can make a fast escape. Again he dispatches a reconnaissance party, who very soon find they are among cannibal giants. Like the Cyclops, but much more accurately since they are not blinded, these monsters also hurl rocks and destroy every ship but one, then carry off the sailors to eat. Only Odysseus and his crew of forty-four save their skins.

Now he heads east and ends up at the island of Aeaea, perhaps at the top of the Adriatic. There he has his encounter with the sea-witch Circe, who first turns half his remaining men, very aptly, into swine. His surviving sailors are terrified, but Odysseus goes bravely, singlehanded, for once Gary Cooper at high noon, to face Circe. On his way to her castle he meets the messenger-god Hermes, who gives him the magical herb *moly*, with scented white flowers and black root, which alone can defeat the enchantress's magic. Robert Graves plausibly suggests the *moly* is the cyclamen. Its best-known herbal effect is aphrodisiac. At first sight Hermes appears rather mysteriously in the story, since it is Athene who has Odysseus in charge. But he was not only the messenger-god. He was also the patron of all thieves and raiders; and even more importantly, a phallic fertility god. One of his 'children' was the Priapus of every ancient orchard, penis erect; another was Hermaphroditus, the homosexual man-woman.

What is interesting at Aeaea is the association with pigs, for if the seal and the dog must head the list of animals hated by sailors, the pig comes a close third, and from far back in history – when it was more often met wild than domesticated. The wild boar would have been just as much a forest hazard as the wolf; and a far worse enemy to crops. Its anger made it sacred (if you can't beat them, worship them) in primitive times, and a relic of this, not just a wise precaution against trichiniasis, lies behind the religious bans on pork-eating. The pig was above all associated with thunder and lightning and high winds, perhaps because of its powers of scenting the unseen . . . such as truffles and dormant cyclamen bulbs (whence the old name of sowbread for the plant). Malevolent spectral pigs are almost as common in British folklore as their canine equivalents. The Isle of Man has a particularly rich selection of them. And well into modern times 'pig' was a taboo word aboard ship; everything to do with the living animal threatened bad luck and bad weather.

The name Circe comes from the Greek *kirkos*, from which we take, almost unchanged, our own 'circus'. The word was used of a wheeling falcon, possibly the peregrine, but the basic meaning is of 'circle'. A related verb means 'to hoop in, to secure by rings'; and that seems to be what lies behind the witch, though her ability to lead by the nose, or snout, lay less in literal rings than her skill with herbs and drugs. She is the patron goddess of all hash-peddlers; yet perhaps also a wish-fulfilment for sailors, in the sense that she could tame the wind-bringing pig.

There is another very interesting sailors' superstition – that a ring of gold is an amulet against drowning. The wearing of it by seamen in a pierced earlobe has now been downgraded to a picturesque costume touch in Hollywood pirate movies; but it once carried much more than a decorative meaning. The gold ring was also worn, by non-seamen, as a cure for short-sightedness. The two superstitions clearly spring from a same notion: that of being able to control fate and natural danger by foreseeing it. This ring can be seen in one very famous Elizabethan ear, which suggests that its wearer was either myopic or had at some point in his life been to sea; that is, if the Chandos portrait really is of Shakespeare. It may also, it must be said, be merely a ring for the ear-jewel beloved of Elizabethan dandies, such as an actor might need professionally; but it is painted with a curious brilliance and conspicuousness, and the young playwright gave very clear evidence that the idea of drowning disturbed him emotionally. The prophylactic ring can be seen in Sir John Hawkins's ear, in Henry Holland's *Heroologia Anglica* of 1620.

At any rate, at Aeaea there is indeed a very sudden change in Circe's character, confirming that she is very far from being hostile to sailors, if properly mollified – or molyfied. The dangerous witch turns melting woman in love; from being a Phorcid, a dangerous siren, she becomes a Nereid, or beneficent one. Odysseus must bed her 'so that in love and sleep we may learn to trust one another' – another gentle and suspiciously unmasculine touch in the telling. Odysseus is bathed, oiled, clothed, fed. His grunting crew are restored to human shape and looked after. Odysseus then falls into Circe's slender arms – and spends a whole year in them. It is another version of lotus-eating.

At last his men force him to move on. Circe, now tamed and gentle, lets her lover go, but tells him he must visit the land of the dead and consult the seer Tiresias before he sails home. He does as she says. The description here is of a far northern land, perhaps a Phoenician tin-trade account of the

Scillies or Cornwall, since it is also a place of perpetual cloud and mist. Odysseus meets Tiresias, who tells him that although he will be allowed home, his voyaging will not be ended. He must set out again and find the people who 'know nothing of the sea'. He also meets, in this Ultima Thule and deepest heart of his journey, his mother and a congregation of the famous dead, the women first. It is clear they take precedence in Hades over the men.

Odysseus sails back to Circe's island. Still his friend, she warns him of his next batch of ordeals, which now go beyond those caused by direct lust or temptation to greed, but are inherent in the nature of sea-voyaging. The first is to sail past the Wandering Rocks (perhaps icebergs) and then past the Sirens, who sit in a sea-meadow (or seal-meadow?) 'piled high with the skeletons of men'. They manage the Sirens only because Circe has told them how to: the crew by plugging their ears with beeswax, Odysseus by tying himself to the mast – so that he can at least hear their song. When he does, he cries to be unbound, a death-wish; but the warned crew only tie him more tightly.

The ship then runs the gauntlet of Scylla the rock-monster and Charybdis the whirlpool. Odysseus steers too close to the terrible Scylla's cavern. Her six heads whip out and snatch six of his sailors from the deck. He does not stop to rescue them. Now his original Ithacan expedition of five hundred or more men is down to thirty-eight.

These survivors come next to the sun-god Hyperion's island, Thinacie. Circe had warned Odysseus that he must not touch the sun-god's flocks, that he had better sail straight on; but his crew virtually mutiny and insist they land for the night. A huge gale blows up, and they are cooped on Thinacie for a month. The inevitable happens: one day they kill some cattle and sheep . . . they are starving by then, so it is no longer greed, but a trap set by the gods. Hyperion goes in outrage to higher quarters, and as soon as the ship is at sea again and out of sight of land, Zeus and a thunderbolt strike. The entire remaining crew are drowned. Now only Odysseus survives, riding the mast and keel. He is washed back to Charybdis, and is nearly sucked down that; then drifts clear. For nine days he drifts, then on the night of the tenth he comes to the 'sinister' western island of Ogygia . . . and to the most enigmatic and, to my mind, most touching of his adventures.

All the above, in the cunning structure of the *Odyssey*, is told in flashback. The novel actually begins with Odysseus on Ogygia. Ogygia may be related to the word 'ocean'; it carried a connotation to the Greeks of great antiquity, primevality. But the island was also highly sinister because it lay in the

west, perhaps in the Atlantic. I suppose that today the west, because of the association with summer holidays (or in America, with the notion of new frontiers and California), bears a generally pleasant sense. It was not so as recently as Elizabethan times, when the west wind was the evil wind, the bringer of storm and disease; and it was even less so in ancient times. The best trading and colonizing opportunities undoubtedly lay westward for the East Mediterranean peoples; but the very word for 'west' in Greek, *skaios*, has an evil, threatening sound. The ancient Egyptians associated the direction with death. Greek ornithomancers faced north to ply their craft and good bird omens always passed on the right, or eastwards. This explains the antediluvian *deiseil*, the Mithraic equivalent of the Christian's sign of the cross before some difficult enterprise: the sunwise or righthand turn. Ships used to make it, sometimes three turns, before a long voyage. All journeys 'to the left' were inherently dangerous.

Athene makes this very clear when she delivers a report on Odysseus during a cabinet meeting on Olympus.

> The island is well-wooded and a goddess lives there, the child of the malevolent Atlas, who knows the sea in all its depths . . . it is this wizard's daughter who is keeping the unhappy man from home in spite of all his tears. Day after day she does her best to banish Ithaca from his memory with false and flattering words; and Odysseus, who would give anything for the mere sight of the smoke rising up from his own land, can only yearn for death.

This, of course, represents Athene-Penelope's official view of the wicked girl, whose name is Calypso. Odysseus backs it when he comes to tell the story of his time on Ogygia later in the text. Calypso, he says, was 'wily'; 'never for a moment did she win my heart.' But then immediately plunges on: 'Seven years without a break I stayed . . .'. Now this is far longer than he stayed anywhere else – even the voluptuous Circe rated only one year – and indeed accounts for over a third of his two decades of wandering.

Furthermore, between these two unkind reports on the girl in Book I and Book VII, we have actually met her in Book V, and Homer gives us a rather different story. Athene has nagged at her father again and Hermes is dispatched to Ogygia to tell Calypso she must give Odysseus up – the gods have further plans for him. Taking the form of what sounds suspiciously like a gannet for the journey (further evidence that the island is either in the extreme Western Mediterranean or the Atlantic itself) he steps on shore 'from the blue waters' and walks along the shore to the great cave where Calypso lives. What greets his eyes is very similar to what Besant's hero saw when he first visited Armorel's farm on Samson; and if it is meant to turn the reader off, it is singularly unsuccessful.

> A big fire was blazing on the hearth and the scent from burning logs of split juniper and cedar was wafted far across the island. Inside, Calypso was singing in a beautiful voice as she wove at the loom and moved her golden shuttle to and fro. The cave was sheltered by a verdant copse of alders, aspens, and fragrant cypresses, which was the roosting-place of feathered creatures, horned owls and falcons and garrulous choughs, birds of the coast, whose daily business takes them down to the sea. Trailing round the very mouth of the cavern, a garden vine ran riot, with great bunches of ripe grapes; while from four separate but neighbouring springs four crystal rivulets were trained to run this way and that; and in soft meadows on either side the iris and the parsley flourished. It was indeed a spot where even an immortal visitor must pause to gaze in wonder and delight.

But not this immortal visitor. After the usual civilities and a cup of tea (brewed ambrosia) the major god and very minor goddess get down to distinctly barbed business. Let me forsake E. V. Rieu for a moment and put it in more modern, multi-corporation terms.

'How nice to see you,' says Calypso, 'and to know that after all head office hasn't completely forgotten I even exist.'

'My dear girl, if you imagine I'd ever come to a godforsaken place like this of my own free will, you're out of your mind. I've never had a more

boring journey in all my life. You provincials don't realize what a desert you live in.' He looks round and yawns. 'It's this miserable what's-his-name fellow you've taken on. The Old Man has other ideas for him. I'm instructed to tell you to remove your tiny claws. Right?'

Calypso springs to her feet, hands on hips.

'You miserable sods! Just because he's not in the company. And I am. I haven't even hidden it, we're as good as married.' Hermes shrugs, says nothing. 'The sheer gall of it! When everyone knows you all spend your life at head office having affaires and chasing secretaries. You're such hypocrites, you do this all the time. And you needn't think I don't know why. One of those ghastly old female department heads has been nosing around again. They're just jealous.' Hermes examines the backs of his fingernails. Calypso is near tears. 'Look, he would have drowned without me. I rescued him, I nursed him, I fell in love with him. I'm even teaching him how he can apply to join the company.' Hermes raises his eyebrows. Calypso stares, sighs, at last surrenders. 'All right. But the Old Man can damn well find the transport himself. I'm not going to.'

Yes, I am vulgarizing a sacred text; but not travestying a very attractive touch of mutinous hurtness in Calypso during that exchange. Hermes leaves. She goes out to find Odysseus moping as usual on the shore, staring out to sea, and realizes her cause is lost just as much with him as at Olympus; and there and then, like Circe, she decides to give him up with gentleness and good grace. She will help him build a boat, provision it for him, send him a fair wind to start out with. The ungrateful man is immediately suspicious: there must be some trick. Will she swear by Styx (the one oath even the gods could not break) that it is all aboveboard?

She tells him, under a half-teasing mask, one or two much-needed home truths then: that he lets his cunning mind rule his human heart; that he ought to know pity can be greater than sexual desire, that the truest love can sacrifice its own existence. She does swear by Styx, but turns quickly away. Homer says that Odysseus walked after her. Let us hope that it was, for once in his life, to make an apology.

Calypso makes one last attempt, at supper that night. She warns him more suffering is to come if he leaves; and promises that if he stays with her, he will gain immortality. They can live together till the end of time. And finally she asks how he can keep thinking of the aging Penelope with a warm young goddess at his side. Odysseus is diplomatic and runs down his wife's looks. True, she's only mortal, but the sea calls, and as for the suffering . . . 'Let this new disaster come. It only makes one more.'

It grows dark, and they have one last night of love. The next morning he is allowed to start building his escape boat.

The Calypso interlude is one of the most endearing in the whole *Odyssey*: a conflict between a dream and a reality, a case of a lonely woman hopelessly in love with a lonely man helplessly in love with his own destiny. It is also one of the most striking cases in all literature of humanity in the writer overcoming the inhumanity of convention. Calypso (whose name holds the *skl* in cipher) should by all the rules of myth be evil. The 'good' characters in the story report her as evil, the hero despises her sexually, even her fellow-gods despise her . . . I remember, the very first time I read the *Odyssey* as a schoolboy, hating Odysseus for leaving her. After many re-readings, despite my knowing that both inner and outer logic must make him leave, I still cannot quite forgive him; and that forked feeling, I am convinced, was not created by a man. At any rate I know how much I owe, as a writer of fiction, to the Calypso-Penelope dilemma; it has haunted my own and countless other novels, and always will.

Odysseus heads for Phaeacia, or Skeria, the modern Corfu. But Poseidon, furious that Olympus has relented (if not the authoress furious that her hero forced her to write that he left Calypso's island 'with a happy heart') dismasts the boat in a violent squall. It drifts in the gale. Then another monster wave sinks the boat. Odysseus strips and swims for it. On the third day he comes to the coast of Corfu, but there is a huge surf, nothing but cliffs. He is carried on to the rocks. 'He clung there groaning while the great wave washed by. But no sooner had he escaped its fury than it struck him once more with the full force of its backward rush and flung him far out to sea. Pieces of skin stripped from his sturdy hands were left sticking to the crag, thick as the pebbles that stick to the suckers of a squid when he is torn from his hole.'

Once more Athene comes to his aid and helps the exhausted man swim along the coast to the sandy cove at a river's mouth; and there at last he can drag himself ashore. He covers himself in leaves under an olive, and sleeps. The next morning the daughter of the King of Skeria-Corfu, Nausicaa, comes with her maids to wash clothes at the river. They play with a ball, and Odysseus wakes. Quick as ever to find his feet, he seizes a branch to cover his nakedness and steps out with a flowery speech to the beautiful princess. Yet another affaire seems about to begin; but this time Odysseus is home. He is befriended by the king and queen and lent a ship to take him to Ithaca to execute his bloodthirsty revenge. Though even there, since he is in disguise to begin with, the odysseys do not cease: he keeps inventing new ones as a 'cover' for his presence, stories of Egypt and the Phoenicians. His old friend, the shepherd Eumaeus, tells another: a king's son by birth, his life too was ruined by the sea and by piracy.

Everywhere, then, the cruel, separating sea; and the folly of sacrificing all to it, when the only tangible and endurable Calypso-Circe-Athene is the one the sailor left behind in the first place, on his own safe home-island and kingdom of Ithaca.

I used in my own first novel – written, like all stories of its kind, under the vast aegis of the *Odyssey* – a famous quotation from T. S. Eliot:

> We shall not cease from exploration
> And the end of all our exploring
> Will be to arrive where we started
> And know the place for the first time.

This does not happen to Odysseus when he at last lands on Ithaca, by a grim irony at the cave of Phorcys, one of Scylla's putative fathers. He fails to recognize his birthplace, partly because Athene has thrown a mist over where he finds himself. In fact he is plunged in gloom. He doesn't know what the people will be like, he doesn't know where to hide the presents that Nausicaa's parents have given him. He wishes he had never left Skeria-Corfu. He has obviously been tricked and marooned on some desert island. His first and supremely typical positive action is to check that none of the presents has been stolen by the Corfiot crew during the landing. (They have left, only to be turned into a reef on their homeward trip by Poseidon, in one last fling of rage.) He starts to weep on the barren shore. Only then does a handsome young shepherd – Athene in disguise once more – appear and tell him where he truly is.

But that first anticlimax, like the curious hesitation on Penelope's side when she does at last recognize him, before the emotion comes, is enormously shrewd; and it does, I think, put a vital accent on the hopelessness of Odysseus' case; on his incapacity to do anything but undergo the experience, the turning of the wheel, even though it finally comes to rest exactly at the point where it started. Odysseus may be wiser, but dire and insatiate he is still condemned to sail on and on, round and round, from island to island, from experience to experience.

The only land where people 'know nothing of the sea' is death; and for better or for worse, the only answer to the mysteries of life lies in the voyage to the islands. In that long penultimate passage of the greatest novel, and greatest homage to the *Odyssey*, of our own century another sailor, Leopold Bloom, is put to the dry question on his own return to Ithaca, at the end of his Dublin day. This is how its crux runs:

> Would the departed never nowhere nohow reappear?
> Ever he would wander, selfcompelled, to the extreme limit of his

cometary orbit, beyond the fixed stars and variable suns and telescopic planets, astronomical waifs and strays, to the extreme boundary of space, passing from land to land, among peoples, amid events. Somewhere imperceptibly he would hear and somehow reluctantly, suncompelled, obey the summons of recall. Whence, disappearing from the constellation of the Northern Crown he would somehow reappear reborn above delta in the constellation of Cassiopeia and after incalculable eons of peregrination return an estranged avenger, a wreaker of justice on malefactors, a dark crusader, a sleeper awakened, with financial resources (by supposition) surpassing those of Rothschild or of the silver king.

What would render such return irrational?

An unsatisfactory equation between an exodus and return in time through reversible space and an exodus and return in space through irreversible time.

What play of forces, inducing inertia, rendered departure undesirable?

The lateness of the hour, rendering procrastinatory: the obscurity of the night, rendering invisible: the uncertainty of thoroughfares, rendering perilous: the necessity for repose, obviating movement: the proximity of an occupied bed, obviating research: the anticipation of warmth (human) tempered with coolness (linen), obviating desire and rendering desirable: the statue of Narcissus, sound without echo, desired desire.

That is Odysseus: the voyage in the mind. The real Ulysses is whoever wrote the *Odyssey*, is Joyce, is every artist who sets off into the unknown of his own unconscious and knows he must run the gauntlet of the island reefs, the monsters, the sirens, the Calypsos and the Circes, with only a very dim faith that an Athene is somewhere there to help and a wise Penelope waiting at the end. No recurrent symbolism in the *Odyssey* is more pertinent than the long and deliberate stripping its hero undergoes: of his ships, of his men, of his hopes, of his clothes, even of his very skin on the cliffs of Corfu. Perhaps the only hope of self-escape for 'the statue of Narcissus, sound without echo, desired desire' lies in that *moly* bloom Hermes hands the sailor at Circe's door; and which James Joyce placed (*shall I wear a white rose*) at the very end of his mistress-piece in his own Mollie Bloom: *yes he said I was a flower of the mountain yes so are we flowers all a woman's body yes that was one true thing he said in his life and the sun shines for you today.*

I should like now to tell the story of a much later real-life Odysseus and his crew, and of the fortunate islands that they discovered. In view of the treasure their voyage eventually brought to light, I am delighted (less now as a novelist than as a local patriot) that the tale begins very close indeed to where I write: to be precise, in Lyme Regis, during bad Queen Mary's reign. In 1554 the wife of a tradesman, John Somers, gave birth to a fourth son, who on April 24th was christened George. A decade later another citizen of Lyme, John Jourdain, also had a son, christened Silvester.

George Somers went early to sea and by the 1590s had become a typical buccaneering Elizabethan captain with many Atlantic voyages and beard-singeing exploits to his credit. Thomas Fuller in his *Worthies of England* reports that he was 'a lamb on the land . . . a lion at sea'; at least part of his prowess was due to his excellence as a navigator. But by 1600 Somers seems to have settled for the lamb's side of it and retired with his laurels, and loot, to Lyme and his wife. She was a local girl, Jane or Joan Haywood, and they were married in 1582, when she was eighteen. In 1603 he became a Member of Parliament for the town and was knighted. In 1604 he was elected mayor. But evidently, like a true Odysseus, he could not just live by the sea, he had to sail it.

In 1609 Somers, a founder member of the London or South Virginia Company, was appointed the admiral of the fleet that was to take a new injection of settlers to the troubled colony. On April 23rd he made his will, and a few weeks later sailed in the accurately named *Sea Adventure* or *Sea Venture* (300 tons), with nine other ships. The sailors and settlers were not only Ithacan in number, about 500, but the settlers at least sound identical in spirit, being (there were also women on board) mostly 'youths of a most lewd and bad condition'. The majority took one look at Virginia and returned to the fleshpots of England at the first opportunity. On board the *Sea Venture* were Sir George's nephew Matthew, Silvester Jourdain, and very probably a number of other Lyme seamen.

However, long before they reached America, the *Sea Venture* had parted company with its little flock. I will put the story in the capable hands of Lyme Regis's first historian, George Roberts.

On the 25th July, the admiral's ship, with the other commanders and their commission, and 150 men, parted company in the tail of a hurricane. The ship worked so much, and became so leaky, that the water rose in the hold above two tiers of hogs-heads. With all hands baling and pumping for three days and nights without intermission, still the water seemed to increase. At last, all being spent with labour, and seeing no hope, they resolved to shut down the hatches. In this extremity, those who had 'comfortable waters' drank to one another as taking their last leaves, till a more joyful and happy meeting in the other world. Sir G. Summers, the skilful seaman, sat all this time on the poop, scarce allowing himself leisure to eat or sleep, steering the ship to keep her upright, or she must have foundered. He unexpectedly descried land: upon the news all ran up, and from ceasing to bale nearly caused their destruction. They spread all sail, though they knew the land to be Bermudas – the land of devils and spirits, then dreaded and shunned by all men. The ship soon struck upon a rock, but a surge of the sea cast her off, and so from one to another, till she was most luckily thrown up between two, as upright as if she had been on the stocks. The wind having calmed, they got out, people, goods, and provisions, in their boats, and arrived in safety without the loss of a man; though some say a league, others half a mile from the shore.

Fallen into a land of plenty and pleasantness, the strangers were lavish in their praise of it. Sir George Summers, like another Aeneas, procured food for the whole company by catching fish with hook and line. They killed thirty-two hogs, which abounded there, said to have swam ashore from a Spanish ship, called the Bermudas, which was carrying hogs to the West Indies.

The Bermudas were first sighted, before 1515, by the Spanish sailor Juan de Bermudez. They were first named Virginiola by the shipwrecked men, then Somers or Summer Islands (the latter probably because of the mild climate, but perhaps also because 'Summer' is a common spelling of Somers in the Lyme documents of the time); and only later by their present name.

The Englishmen, as Englishmen will, attributed their good fortune to the fact that they were English; the island's malevolence clearly extended only to wicked Catholic foreigners like the Spaniards and the French, though they were at first puzzled by mysterious noises at night and worried, as superstitious seamen might well be, by the ubiquitous pigs. However, they quickly developed a great fancy for their enforced home, despite its evil reputation – in fact behaved exactly like Odysseus and his crew on Circe's island of Aeaea after that first little contretemps, also to do with pigs. The climate was delicious, there was wood and fresh water, palm-leaves for roofing and walling, sea-fowl (apparently petrels or shearwaters) 'full and fat as a partridge', turtles, fish 'dainty as salmon'; and the pork had 'more pleasant and sweet a taste than mutton in England'. Even the Bermuda crow had 'as white flesh as a chicken'. Readers of contemporary travel-agency literature may notice a certain familiar ring ('superb sea-food, endless unspoilt beaches') in these similes, and they will be quite right. Most of the gentlemen reporters on all the early American 'ventures' had a heavy financial stake in their success; if they did not quite yet show the monstrous blind eye exhibited by some Victorian emigrant-recruiters in Europe, they were very decidedly not interested in turning customers away.

The idyllic side of this first involuntary holiday in the Bermudas was short-lived. The soldiers, sailors and settlers of the *Sea Venture* had scattered among the various islands, and quarrels and mutinies soon developed. Greed was at work again; there was a plot to murder Somers and seize the stores, which was scotched. The sea-courage of the Elizabethans is inextricably ravelled with personal ambition and litigiousness. In 1588, for instance, the whole of Lyme was up in arms – not, as romantic tradition would like us to believe, against the Spanish Armada, but against the neighbouring town of Axminster for failing to cough up its share of the Armada levy. There is barely a mention of the glorious victory, only a series of outraged squawks to the Privy Council about who ought to pay what. 'A greater shipwreck,' wrote William Strachey of this same expedition to Virginia, was made by 'the tempest of dissension: every man overvaluing his own worth, would be a commander: every man underprizing another's value, denied to be commanded . . . every man sharked for his present booty.'

In the first escape attempt, fourteen men set out in one of the *Sea Venture*'s boats for the American mainland. They were never heard of again. Somers then built two small boats, probably using a mixture of the *Sea Venture*'s timbers and the local juniper (Bermuda 'cedar'). On May 10th, 1610, the two pinnaces set out for Jamestown. They made the six hundred mile passage in only thirteen days, thanks once again to Somers's seamanship and expertise in navigation; but only to find the rest of the original expedition much less happy with their unshipwrecked lot. They were starving, and they had Indian troubles.

Somers eventually agreed to sail back to the Bermudas in company with Samuel Argall, the later kidnapper of Pocahontas, to fetch meat and fish for these less than brave new-worlders. He was separated from Argall, but he arrived back in Bermuda in early November – only to die there, on November 9th. The cause of death was 'a surfeit of pig'. His last order to his nephew Matthew was to take a cargo of the 'black hogs' back to Jamestown. Perhaps the crew mutinied at the thought of taking live pigs on board and once more turning their backs on home; perhaps the new captain decided to have a grim revenge on their superstitiousness. At any rate, having buried his uncle's heart and entrails in the islands, he secretly sealed up the presumably well-salted corpse in a juniper box, smuggled it aboard and set sail for Lyme.

The ship did the voyage safely, in spite of its dark cargo. Sir George's remains, carried 'athwart and first ashore' if tradition was obeyed, were interred on June 4th, 1611, with full military honours; and lie to this day beneath the vestry floor in the church of Whitchurch Canonicorum. Somers had a manor-farm in the parish, on a hill overlooking the sea and his birthplace. His wife, I am afraid, proved no Penelope. An entry in the Whitchurch parish register records that on July 12th, 1612, 'Lady Sumers' married a certain 'William Raymond, Esquire' – no doubt a buccaneer of a different, safer kind . . . or perhaps a fool. When the will was proved in November of that year, it turned out that Matthew inherited all his uncle's considerable real estate. It was evidently not quite pure piety that made him risk Scylla's fury by bringing that indisputably dead body back.

Silvester Jourdain had meanwhile sailed in another ship straight back to England from Virginia. It carried an official dispatch to the Company patentees, which was to be rewritten and published later that year as *A True Declaration of the Estate of the Colony in Virginia*. This was drafted by William Strachey, another of the Bermuda survivors, who had also written a private – and much more truthful – account from Virginia in a letter dated July 10th, 1610, though it was not published until 1625. But Jourdain evidently sniffed a scoop to be made; or perhaps wanted to play the mini-Homer. As soon as he was back he rushed out a pamphlet entitled *A Discovery of the Barmudas, otherwise called the Isle of Divels*; and this was the first publicly available account of the extraordinary adventure.

I can add a few details to the meagre and not always accurate account of him given in the *Dictionary of National Biography*. Silvester Jourdain (more often spelt Jurden or plain Jordan) was baptized on February 14th, 1565. He was the third child, having an elder brother and sister, and at least three younger siblings. His father was not William Jourdain, as in the *Dictionary*, but John. William did have a famous son, Silvester's cousin Ignatius (1561–1640), the Puritan preacher and staunchly independent later mayor of Exeter. ('Are you not afraid of my Lord Keeper, Mr Jourdain?' 'No, sir. The Lord is my Keeper.') Another cousin, also John Jourdain, was an important figure in the early East India Company. He died in a sea-fight in 1619.

The Jourdains (Silvester's father had been mayor in 1577 and 1584) were plainly an enterprising and well-circumstanced family. Silvester was a free burgess of Lyme, but no staid pillar of the community. He was several times complaining or complained against in the Lyme court records of the 1590s. In 1594 he was charged with blocking a right of way; in 1596 fined 12 pence for playing at bowls (on the same occasion one of the Somers family was fined 9 pence for brawling and drawing blood); in 1597 he was indignant

that the 'schoolmaster which teacheth grammar' allowed the boys to use his garden as a lavatory; and in 1598 he was condemned to five hours' confinement by the town council for contumacious behaviour – he would not stand for it, and forfeited his civic freedom by decamping. In 1612 he returned to the Bermudas, in a voyage that took two and a half months; and produced another pamphlet, *Newes from the Bermudas*. He never married, and died in London in 1650, at the ripe old age of eighty-five.

Quite apart from his anxiety to get his pamphlet out, nothing is more probable than that Jourdain would have been in London in 1610 to tell the story to the many backers of the Virginia adventure. One person in particular who would have wanted to question him was the Earl of Southampton, since he had patronized the Weymouth and Harlow voyages to Virginia of earlier in that decade, and was another founder of the Company; and in the Southampton circle was a sharp-eared and myth-prone playwright with as good a nose for the topical as that of Sir George Somers for magnetic north. He was also no mini-Homer, although he had had some trouble adapting his gifts to the new fashion for the pastoral, a form primarily concerned with the contrast between nature and culture . . . the debit-and-credit of human progress and civilization. Like all men of his time, he had had a long love-hate relationship with the symbolism of the sea-voyage, and a particular obsession with death by water. It had first declared itself nearly twenty years before, in one of his earliest plays, and he must, in the winter of 1610, have remembered the relevant passage. A man recalls a nightmare.

> Lord, Lord! methought, what pain it was to drown!
> What dreadful noise of waters in mine ears!
> What ugly sights of death within mine eyes!
> Methought I saw a thousand fearful wracks;
> Ten thousand men that fishes gnaw'd upon;
> Wedges of gold, great anchors, heaps of pearl,
> Inestimable stones, unvalued jewels.
> All scatt'red in the bottom of the sea:
> Some lay in dead men's skulls . . .

The man goes on to explain how in the dream he could neither die nor wake.

> O, then began the tempest to my soul!
> I pass'd, methought, the melancholy flood,
> With that grim ferryman which poets write of,
> Unto the kingdom of perpetual night.

All through history literary scholars have searched externally for sources; which is not necessarily wrong, but overlooks one very simple fact that any practising author could tell them. The major influence on any mature writer is always his own past work. The tyro dramatist from Stratford may, in the lines above from *King Richard the Third*, superficially have been trying to out-Marlowe Marlowe; but he was also sowing a seed for future germination. He knew the *Aeneid* much better than the *Odyssey*, but Odysseus' experience already haunts this passage: the onward hubris and vaulting ambition of the voyage, the black stasis of shipwreck and drowning, the gateway beyond of ultimate revelation. Clarence's dream stayed dry tinder, waiting for the Bermudan spark.

I must not let local patriotism run away with me. In fact Shakespeare seems to have taken rather more from Strachey's letter, which he must have been shown, than from Jourdain. It is not at all unlikely that he met one, or even both, of them. But the key figure in the story is really the man whose bones lie under the vestry floor; who kept the *Sea Venture* afloat, who maintained some sort of order during the difficult winter on the islands, who organized the escape and brought it and the two writers he had aboard to a successful conclusion. I think it is to him, the only begetter in non-literary reality, that we most owe the one other exploration of the island metaphor that stands shoulder to shoulder with the *Odyssey* and *Robinson Crusoe*.

If *The Tempest* is ever filmed, I know exactly where its first scene must be shot. It is at the northern end of Bryher, at Hell Bay, where the Atlantic in a rage can be at its most awesome and terrible. The huge white lasher that soars off Shipman Head then is intimidating even from St Mary's, four miles away. *Now would I give a thousand furlongs of sea for an acre of barren ground – long heath, broom, furze, anything . . .*

Like so many geniuses of the first order, Shakespeare seems to me to have saved his profoundest work to the very end. *The Tempest* (its first known performance, before the king, took place on November 1st, 1611) may not have the greatest poetry, or the greatest penetration of human character. Its brevity, it is the shortest but one of all his plays, its enormous compression in time and space, its cuts, leaps and discontinuities can make it seem lightweight, no more than a sketch. But its lightweightness is that of Cézanne's last water-colours or of the *arietta* from Beethoven's last piano sonata – whatever such art may lack in substance it gains, like a sublime thistledown, in altitude. The brusquenesses, even the clumsinesses, are those of supreme mastery, of a man sailed far beyond the barrier of mere technique; to where those lines I quoted from T. S. Eliot gain their greatest force and justification.

The Tempest floats free in a way that no other of that formidable chain of masterpieces of the seventeenth-century years of Shakespeare's life quite manages. They are for the world at large. It is for each; and that is why Shakespeare made his over-riding metaphors the island and the sailor stranded in a place that he cannot fully understand . . . that both bewitches and is intensely cruel, that can hold both Calibans and Ariels, Antonios and Mirandas, that can be only too savagely 'real' and yet still an insubstantial pageant. Of course Prospero's island lies no more in the Bermudas than it is, according to the story, set between Tunis and Italy – though that latter location strongly echoes those of both the *Aeneid* and the *Odyssey*. *The Tempest* is a parable about the human imagination, and thus finally about Shakespeare's view of his own imagination: its powers, its hopes, its limits – above all, its limits. The play's true island is our planet, in its oceanic sea of space.

Any specific and realistic shape the island location took in Shakespeare's consciousness must derive from his knowledge of the Strachey, Jourdain and associated pamphlets, and I am not for a moment proposing the Scillies as an alternative; yet they were far more on Elizabethan and Jacobean minds than they are on ours today. This was because of their vital strategic importance during the chronic Armada scares of the period – which by no means began or ended with the debacle of 1588. In August 1601, the mayor of Lyme was urgently ordered 'to set forth a barque . . . for the discovery of

the Spanish fleet'; there was a major panic as late as 1628, when many Scillonians fled to the mainland. Owing to the difficulty of sailing in convoy, the Spaniards were well aware that they needed an offshore rendezvous before launching a final attack, and the Scillies were the obvious place for it. The problems of fortifying the islands (first undertaken in 1548), and of exploiting their value as an early warning system, crop up continually in the State Papers of the period. The very notion of an ordained ruler having the would-be usurper wrecked and brought to justice would have remained highly attractive to anyone who had lived through the worst years of the Spanish threat. But there is something else on the Scillies that does bring us, symbolically, much closer to the play.

Two summers ago I spent a few days on one of the least-known and most beautiful of the larger European islands, the queen of the Baltic, Gotland. By pure chance one morning, out walking, I came through some fir trees on the edge of a remote strand on its east coast and stumbled on something that took me immediately back to the Scillies: a maze of beach-pebbles laid in concentric rings, as if by some playful group of idle teenagers. But I knew I was staring down at something much more ancient and haunting than that. There is just such a maze on the western side of St Agnes, also standing on a little slope above the sea, and looking out towards the ships' graveyard and seal-city of the Western Rocks.

It is near a farm called Troytown; but that name comes from the maze itself. 'Troy-fair' and 'Troy-town' are very old dialect words meaning a mess, a confusion . . . a maze. These particular pebble mazes, usually of ten to fifteen yards in diameter, are mostly found in Scandinavia, where they have a very close association with coasts and islands. According to Geoffrey Grigson they usually carry similar names there: Trojeburg, Tröborg, and so on. There is another famous one on Gotland, at Visby. R. L. Bowley says the St Agnes maze has been recorded for two hundred years and was 'probably originally constructed by a bored lighthouse-keeper'; but I think, like Geoffrey Grigson, that the evidence is very much against this. We know the Vikings knew Scilly, and the similarities with indisputably much older sea-mazes in Scandinavia are too great.

We shall never know what ritual significance these mazes had to the Vikings, but in both Celtic and Mediterranean Europe the maze appears to be associated with the tomb; and escape from it, with reincarnation. This is what lies at the heart of the Daedalus legend: the real labyrinth he escaped from in Crete was the maze-pattern of the very ancient spring-fertility or 'partridge' dance (more accurately the migratory and cornfield-haunting quail, still a prime harbinger of summer on every Aegean island). It certainly pre-dated Minoan Crete, and was probably originally performed on literal threshing-floors in the spring, and only later on the symbolic threshing-floor of the maze.

I have an even older fragment of antiquity standing beside my statuette of Gilgamesh triumphant over hostile nature – a fat old pot from the third millennium B.C., excavated in one of the mountain plains just north of ancient Sumeria. In a kind of strip-cartoon round its shoulder stylized two-headed birds peck and bob in a field of corn; and among them one can see the most famous of Minoan-Cretan symbols, the *labrys*, or so-called double-headed axe. But the two triangles of which it is formed are in fact simply even more stylized headless birds. Another highly stylized symbol on the pot, of four inward-pointing triangles, is related: this stands for four deer,

or cattle, running round a pool. A case in the British Museum is devoted to the elaboration of this design.

The *labrys* that supposedly guarded, or warned against entering, the labyrinth is no axe; but a dancing-bird symbol of fertility, or food. *Labrys* is not a Greek word, but I believe its origin can be plausibly guessed. One has only to say *labr-* to feel the mouth and lips move to peck in and gulp. There is a Greek adjective *labros* that means 'forceful and greedy' – 'gluttonous' by extension. Another word, *labbax*, meant a sea-wolf. *Labrum* is Latin for 'lip'; and their *labor*, our labour, is based on an ancient sense of slipping away and of the consequent suffering and anxiety (which we retain in the 'labour' of childbirth). The need to eat, the need to work to eat, the need to propitiate the forces that control fertility and climatic conditions . . . this is what the ancient labyrinth or maze truly signifies. Even the monster at its Cretan centre, the bull-man Minotaur, is a fertility symbol. I know the maze on St Agnes was first built not by a bored lighthouse-keeper of the eighteenth century, but by a Phoenician sailor two and a half thousand years earlier; and know equally well that no serious archaeologist would for a moment support such a hypothesis.

The maze is also a very ancient symbol of ingenuity in craftsmanship, of the ability to fabricate, to sew and weave beyond ordinary skill – in other words, it is the prime proof of the artificer, or artist. If Minos stood for sea-power, exploiting commerce, and Scylla stood for its greatest enemy, hostile nature and the shipwreck, Daedalus stands for the producer who inspires the endless conflict between profit and its loss. Nothing is more poetic, both symbolically and in justice, than the end of the Daedalus-Minos legend.

Minos keeps the great artificer and his son Icarus prisoners on Crete. Daedalus invents flight, but his son goes too close to the sun and is a victim of the first fatal air-crash. Daedalus buries him, then flies on to Italy, and eventually to Sicily, where he works for King Cocalus (another ominous-sounding sea-name). Minos, not one to accept brain-drain, sets out to find his disobedient inventor. He knows the fugitive is in hiding, so once in Sicily he sets a problem he knows only Daedalus can solve: to pass a linen thread through the complex convolute chambers (the maze symbolism again) of a Triton shell. Daedalus is tempted and, by a brilliant piece of what Mr Edward de Bono would call lateral thinking, solves the problem. Minos now knows the scent is very warm. But King Cocalus's daughters warn Daedalus and a plot is made, neatly echoing the fate that Minos once callously left Scylla to meet. He is persuaded to take a bath. Daedalus constructs an ingenious pipe. As soon as the sea-emperor is in the tub, it is flooded with boiling water (or pitch, in another version) and he (Cretan sea-power) is done for. *Omnia vincit ars*; and what are all the Somerses, Southamptons, Jourdains now but skulls and bones in Shakespeare's cellar? It is not Odysseus who finally survives, but Daedalus. 'O, my name for you is the best,' cries Buck Mulligan to Stephen Dedalus at the beginning of *Ulysses*. 'Kinch, the knifeblade.'

All of which may have escaped Shakespeare's small Latin and less Greek; but mazes he would have known. They were much commoner – especially in the unicursal, as opposed to multicursal, form of the one on St Agnes – in the England of his day. They were usually made of turf, not shore-pebbles. In another play drenched with magic, *A Midsummer Night's Dream*, Shakespeare already has Titania lamenting their disappearance. 'Quaint' carries its old sense of 'ingenious' or 'cunning'; not its modern one.

> The nine-men's-morris is fill'd up with mud;
> And the quaint mazes in the wanton green,
> For lack of tread, are undistinguishable.

No doubt it was the Puritans who were historically responsible for the loss, about this time, of the old morris floors and turf-mazes, with their superstitious associations. One explanation of Prospero's final curtain-speech is that it is a tacit apology to King James for the references to magic, witchcraft, mazes and demonology in general. It is not an explanation that satisfies me, although such 'material' may well have been suspect to the pious and conventional of the period. What is certain is that Shakespeare did deliberately plant the maze symbolism in *The Tempest*. The very structure is circular and maze-like, and there are a number of direct references. 'Here's a maze trod indeed,' groans old Gonzalo, 'through forthrights and meanders!' Then Alonso in the last act: 'This is as strange a maze as e'er men trod, And there is in this business more than nature Was ever conduct of.' At the very end, when Gonzalo blesses Ferdinand and Miranda, he adds: 'For it is you that have chalk'd forth the way Which brought us hither.' All of us have found ourselves, he says in conclusion, 'Where no man was his own'.

It must be remembered too that the verb 'amaze', also used at key places in the play, had a far more literal connotation then – of trance, of almost total loss of normal bearings and physical capacity. 'Half sleep, half waking' is how Shakespeare himself glosses 'amazedly' in *A Midsummer Night's Dream*. Men of his time knew where the real monster lay in the labyrinth – not at its centre, but in the difficulties of finding the right path to it. To the more sophisticated Elizabethans and Jacobeans the maze bore a close analogy to the ring-diagram of the Ptolemaic universe (with its anthropo-centrism) and to astrology, where each planet-path symbolized an aspect of psyche; perhaps also to the tortuous search for the philosopher's stone. When Prospero says 'Now does my project gather to a head' near the end of the play, he is certainly using the jargon of alchemy. Maze-centre repre-sented true self-knowledge.

One of the naive cuts in Francis Quarles's *Emblems* of 1635 – perhaps the most popular illustrated book, after Foxe's *Martyrs*, of that soul-searching century – shows this interpretation very clearly, and even the association with the sea. Anima, the pilgrim soul, stands at the maze centre, holding a cord thrown down by an angel on a lighthouse with a burning cresset. At the beginning of the maze, a blind man follows his dog, and just outside it, drowning men raise arms for help; two others try to clamber the rocks on which the lighthouse is set. There are ships on the horizon. It is a very strange picture, since the maze (in this case of the multicursal type) seems set in the sea, with its passages cut in water. A stanza of the accompanying poem explains this bizarre conceit. The 'labyrinth' is the world.

This gyring lab'rinth is betrench'd about
  On either hand with streams of sulph'rous fire,
Streams closely sliding, erring in and out,
  But seeming pleasant to the fond descrier;
  Where if his footsteps trust their own invention,
He falls without redress, and sinks without dimension.

And Gonzalo:

All torment, trouble, wonder and amazement
Inhabits here: some heavenly power guide us
Out of this fearful country.

The proving maze in *The Tempest* is constructed by Shakespeare's imagination, hiding behind the mask of Prospero. At one level it is very similar to our most familiar contemporary use of the word: the laboratory maze for testing learning-ability and behaviourism in animals; and, as has been often pointed out (and like, I am afraid, a good deal of laboratory testing), it does at this level little but underline the obvious. True, two nice young people fall in love, but out of their own natures, not magic. A

Oh that my wayes were directed to
keepe thy ſtatutes. pſal. 119. y.
192.

fuddleheaded but kind old man, Gonzalo, remains fuddleheaded and kind to the end; two cynical, scheming politicians demonstrate by their final bitter silence that they will always be so; two seamen-buffoons and an Indian 'savage' stay unredeemed. Even the spirit-agent Ariel seems anxious to be freed from playing assistant to any more such futile experiments. Only Alonso, who conspired in the usurpation, shows any plausible repentance; but he has very little to lose by changing sides – it is mere shrewd diplomacy, when his son is to marry the heiress to Milan.

It is certainly not difficult to read, even in Prospero himself, a suspicion that he has in vain tried to surpass that other sea-magician and pig-maker who hides behind Caliban's Bermuda-inspired mother Sycorax (Greek *sys*, sow, and *korax*, crow) – Circe. The 'every third thought shall be my grave' he prophesies of his return to Milan is hardly a happy final note. He forgives, as Circe and Calypso forgave their visitor (his last command to Ariel repeats their last gift to Odysseus, the provision of a fair wind), but the forgiveness is also like theirs in its air of forced circumstance, of *noblesse oblige*. No hearts have changed.

I spoke earlier of Homer's humanity overcoming convention in the treatment of Calypso, and I sense something similar in *The Tempest*: a wise sadness seeping into the ritual happy ending. The play may outwardly demonstrate true culture, or moral nobility, triumphing over both false culture and culturelessness; but it throws strange doubts and shadows on its own message and on its very form. The conflict revealed is the oldest in all art, and takes place inside the artist: between the power to imagine and the use of imagining. *Cui bono*, to what purpose? What will it change? The question haunts every constructor of worlds that are not the one that is the case. Solution is not helped by that other secret every artist nurses: all the incommunicable pleasures of maze-construction, of sea-voyaging, of island-discovering, that may be infused in the final product, but are never explicit in it. The truth is that the person who always benefits and learns most from the maze, the voyage, the mysterious island, is the inventor, the traveller, the visitor . . . that is, the artist-artificer himself.

I believe this is precisely what Shakespeare realized during the course of creating *The Tempest*: that a terrible solipsism underlay the play. Who benefits most is the maker of it, Prospero-Shakespeare; and what he learns most is the dubious efficacy of the demonstration to all who merely undergo it, as opposed to the one who designs it. This is one major reason why the maze-running changes so little in the basic natures of the guinea-pigs, and why there is that famous reference to drowning his books and turning his back henceforth on magic. Three times Caliban tells Stephano and Trinculo that Prospero's books must be destroyed before he can be murdered.

Of course this may seem supremely unimportant to the outsider, all of us in the audience: enough that the maker here is a great poet and dramatist. *The Tempest* now has the status of a personal myth become universal; and like all myths, it allows of countless interpretations, maze upon maze, which makes it additionally delicious meat to an age so besotted by analysis and dissection, so devoted to daedalizing Daedalus. The growing tendency of our century is to reify; to put learning above living. It was even a fault in Prospero himself. We sometimes forget how he first lost his dukedom in Milan. Here is his own notoriously stiff and knotted (in syntactical and metrical terms) account to his daughter. His reign in Milan was

> . . . for the liberal arts
> Without a parallel; those being all my study,
> The government I cast upon my brother,
> And to my state grew stranger, being transported
> And rapt in secret studies.

He goes on:

> I, thus neglecting worldly ends, all dedicated
> To closeness and the bettering of my mind
> With that which, but by being so retir'd,
> O'er-priz'd all popular rate . . .
> . . . Me, poor man, my library
> Was dukedom large enough . . .

And the books that had been his downfall were, it will be remembered, secretly smuggled by old Gonzalo on board the boat that took him into forced exile. The 'books' are his imagination, and he cannot be parted from that, whatever the final talk of drowning 'deeper than did ever plummet sound'.

The climax of the self-doubt lies in that very last speech of the play, despite its surface flatness and near-banality. One can take it as the trite request for applause as much expected at the time as the flowery dedications to rich patrons that began most books. But it reads rather differently if we posit that Prospero really stands for Shakespeare's own power to create magical islands of the mind; and that what has gone before (and far beyond *The Tempest* itself) raises very considerable doubts about the ability of that power to change human nature in any but very superficial ways.

Now my charms are all o'erthrown,
And what strength I have's mine own,
Which is most faint: now, 'tis true,
I must be here confin'd by you,
Or sent to Naples. Let me not,
Since I have my dukedom got
And pardon'd the deceiver, dwell
In this bare island by your spell;
But release me from my bands
With the help of your good hands:
Gentle breath of yours my sails
Must fill, or else my project fails,
Which was to please. Now I want
Spirits to enforce, art to enchant;
And my ending is despair,
Unless I be reliev'd by prayer,
Which pierces so, that it assaults
Mercy itself, and frees all faults.
As you from crimes would pardon'd be,
Let your indulgence set me free.

What is striking are the repeated references to imprisonment and release. The illusion, the magic, is over; but the innermost meaning, the conclusion from the maze-heart of the imaginative power, is not. The prison for Shakespeare is in having failed to communicate ('I must be here' – in the marooning artifice of the medium, the stage – 'confin'd by you') that the power to affect, and effect, by imaginative means is strictly dependent on precisely that same *active* energy of imagination in the audience that lay behind the creation. The 'spell' is their literalness, or blindness. What makes the island 'bare', the ending 'despair', is the putting of art in the 'bands' of parenthesis, the treatment of it merely as ingenious maze, external form, surface of text and image, entertainment.

It is almost as if Shakespeare foresaw the very recent neo-Freudian theory of language and literature as the prime alienators of self from reality; a universal exploitation far worse than the social and economic ones. Something castrating haunts both the Greek and the Latin words for 'I write' – *grapho* and *scribo*, which have a shared Indo-European origin meaning 'to cut'. It is seen in all the ancient magical uses of the sign or symbol that forbids or proscribes. The text may demand action; but something in its external, objective, 'other' nature is always inherently alienating of action in all who did not perform the original action of its

creation. In simplest terms, and at even the very highest level, it is someone else's magic.

We are, in that epilogue, aeons beyond the make-believe world of benevolent wizards and cowslip-dwelling sea-nymphs. Only one wind will ever fill those sails, if the barrenness of our planet-island is ever to be truly left behind; and it is beyond even Shakespeare's power to provide it.

But that the vital motive power must lie in the imagination of the beholder is implicit in the piece. It has, I think justly, been interpreted as a play with a cast of one: that is, its eleven main parts can all be seen as aspects of the one mind, so many planets of different humours mazing and circling round a central earth; and the central earth of *The Tempest* is clearly Prospero-Ariel, the imagination and its executor, the writer and his pen, forced to cohabit with all the other dispositions of the mind . . . and all stranded, like a schizoid Crusoe, on the island of the each. *Hamlet* and *Lear* and *Macbeth* are dreams of men. *The Tempest* is a dream of the only divinity men are allowed, and a statement of how literally island-small, compared to the all-controlling power of the gods of myth, it is – and always will be until it is shared and understood.

Every child who visits St Agnes has a hop round the Troytown maze; and I hope will long continue to do so, for I should hate to see it fenced in and museumized like Stonehenge. But to anyone who has lived for longer than a summer holiday on a remote island, I trust it will always mean rather more than a minute's amusement. Most who have been through that experience will know the maze began at their first step ashore; and I should be surprised, even though the centre, given the present world shortage of Prosperos and Ariels, was not reached, if any in retrospect regret the exploration.

I first began to feel the releasing power of *The Tempest* when I lived on my island in Greece – the lack of a Prospero, the need of a Prospero, the desire to play Daedalus. It is the first guidebook anyone should take who is to be an islander; or since we are all islanders of a kind, perhaps the first guidebook, at least to the self-inquiring. More and more we lose the ability to think as poets think, across frontiers and consecrated limits. More and more we think – or are brainwashed into thinking – in terms of verifiable facts, like money, time, personal pleasure, established knowledge. One reason I love islands so much is that of their nature they question such lack of imagination; that properly experienced, they make us stop and think a little: why am I here, what am I about, what is it all about, what has gone wrong?

Modern wreck-divers use the word 'crud', a dialect form of curd, to describe the coagulated minerals that form round any long-sunken metal object, and that have to be laboriously chipped and leached away before it can be exposed to sight again. Islands strip and dissolve the crud of our pretensions and cultural accretions, the Odyssean mask of victim we all wear: I am this because life has made me like this, not because I really want to be like this. There is in all puritanism a violent hostility to all that does not promise personal profit. Of course the definition of profit changes. 'It may be thought,' wrote Cromwell to the House of Commons after he had helped reduce the city of Bristol in 1645, 'that some praises are due to those gallant men, of whose valour so much mention is made: their humble suit to you and all that have an interest in this blessing is that in the remembrance of God's praises they be forgotten.' The profit then lay in eternity and the ticket to that was to be bought by an arrogant extreme of self-denial.

We have in our own century lost all faith in the remembrance of God's praises. The profit now is tallied in personal pleasure; but we remain puritan in our adamant pursuit of it. Purveying recipes for pleasure has become a mainstay of popular publishing and journalism: where to go, what to enjoy, how to enjoy, when to enjoy, to such a clogging, blurring extent that our modern duty to enjoy is nowadays almost as peremptory – and destructive – as the old puritan's need to *renounce* pleasure . . . to ban the play, the dance, the graven image and everything else that makes present life agreeable. It is all very well creating a permissive society. But we have not created the essential corollary of a pagan mind.

The true pagan mind, from Homer on, always knew that the laws of pleasure have very little to do with endless consumption, endless exhortation to experience, endless attempts to tell the individual in what his pleasure consists and to guide him through the labyrinth whose deepest values can only be self-discovered. We cannot all be labyrinth-makers; but we can all learn to explore and trace them for ourselves. There used to be a guide to the famous maze at Hampton Court that showed the quickest route to take. Nobody who used it ever reached the centre; which lies not in the unravelled, but the unravelling.

We have helicopters, motorboats, guided tours; all the facilities, all the knowledge. Shakespeare, despite that gold ring in his ear, very probably never saw a remote sea-island in all his life, so his is merely charming fantasy, over-intricate and increasingly incomprehensible, like his rich language in our tired, etiolated period of it; and he himself merely the world's greatest dramatist (from Greek *drama*, a deed, an acting), safe-throned on the peak of Parnassus, at a very great and alienating distance from you, me and anything else relevant today . . .

But now I am hopping into a sermon, and that would never do. Like all good islands, the Scillies can play their own pulpit. To those who cannot go there, will perhaps never go there, I can at least recommend Fay Godwin's photographs, which represent very exactly what that wise visitor I began with will look for: the elemental compound of sky, sea, sand, rock, the forms and textures of simplest things, the cleansing, as the sea itself will cleanse, of over-artifice, over-knowledge and over-civilization from the mind exhausted by our age's mania for the second-hand, mechanical image. Our century has rightly learnt to admire the Zen gardens of Japan for their austere simplicity; and most of Scilly remains one huge Zen garden of the Atlantic.

A medieval master of that faith was once asked by a novice which plant in a garden most pleased the Buddha. The old man thought, then answered.
'The mirror.'
'But master, a mirror has no leaves, no flower, no fruit.'
My sincere hope is that the slap the novice then received was freighted with each ounce of force still lying in the sage's arm.

The Photographs

*Copyright © John Fowles